A Trip into Time?

You buy a new watch, drive to the airport, and, just before you jump on a plane that will fly you around the world, you synchronize it with another new watch you bought for your mate. A week later you return from your trip and meet your mate at the airport. But when you compare the time on the two watches, yours is running hours slower.

Believe it or not, neither watch is broken. The only thing you need to fix is your understanding of Einstein's theory of special relativity. Find a simple explanation in . . .

D0862911

THE POCKET
PROFESSOR

PHYSICS

THE POCKET
PROFESSOR

Everything You Need to Know About
PHYSICS

Gregg Stebben
and
Daniel Orange, Ph.D.

Series Editor: Denis Boyles

POCKET BOOKS
New York   London   Toronto   Sydney   Tokyo   Singapore

An *Original* Publication of POCKET BOOKS

POCKET BOOKS, a division of Simon & Schuster Inc.
1230 Avenue of the Americas, New York, NY 10020

ISBN: 0-671-53490-4

First Pocket Books trade paperback printing October 1999

10   9   8   7   6   5   4   3   2   1

POCKET and colophon are registered trademarks of
Simon & Schuster Inc.

Book design by Helene Wald Berinsky
Cover design by Tom McKeveny
Illustrations by ARB

Printed in the U.S.A.

RRDH/✖

For Bonnie,
and everyone else who enjoys learning about,
and talking about, science

ACKNOWLEDGMENTS

For their years of fostering my curiosity in science and for never complaining about the aftereffects, I thank my dad and mom, Arnold and the late Elise Orange. My high school physics teacher, Fred Lynn, got me started, and my professors at MIT continued my training, particularly John Southard, Roger Burns, and Frank Spear. Peter Molnar taught me that humanity can be at home in the heart of a scientist. At MIT I found out that you can work as hard as you play, and vice versa, thanks to the Baker House New Year's Crowd. Norman Burtness, now a high school physics teacher, provided thoughtful input to this effort. Another group of great teachers at U.C. Santa Cruz, Bob Garrison, Eli Silver, Rob Coe, and Casey Moore, showed me that it is possible for great scientists to be nice guys, too. For science, adventure, and friendship, thanks to John Carl Adams, Andrew Cohen, Dan Farber, Jeff Marshall, and Jeff Emery. Finally, I want to thank Bonnie Holmer for her love and support, which defy the laws of physics by creating more every day than there was the day before.

—D.O.

Contents

PHYSICS

WHAT GOES UP MUST COME DOWN

Here's some basic science: What goes up . . . must come down.

And everyone knows it. Airline pilots. Center fielders. Window washers. Stockbrokers. Pastry chefs. Parents of small children.

But why must it—whatever it *is—come down?* Figure that one out and you're well on your way toward figuring out this whole physics ball of wax. Here's the thing: Regardless of how much or how little of it you understand, physics is everywhere you look. In fact, you need physics just to explain the very act of looking.

Sure, it's easy enough to walk around thinking you've got a pretty good handle on the world around you. Why does a ball fall when dropped? Gravity. Why does a tree make a sound when it falls in the woods? Sound waves. Why does a light come on when we flip a switch? Electricity.

Unfortunately, knowing the answer to these questions isn't the same as knowing what gravity *is,* what a sound wave *is,* or what electricity *is*—instead, using these one-word retorts just proves you were paying attention that day in class and memorized the answer in case anyone ever asked.

So just for fun, forget all that stuff you memorized in high school and imagine that the natural world is a complete mystery to you. As much a mystery as, say, the stock market is to most. If this was the case, wouldn't you start looking around and noticing a few things and trying to bring some order to the crazy, mixed-up world around you?

WHY DOES THE SUN ALWAYS SET IN THE WEST?

One of the first things you'd probably notice is that giant ball of fire that comes up over one field every morning and goes down over the opposite hill every evening. And if you were smart, the next thing you'd probably ask yourself is "Why does the sun always set in the west?"

You know the answer to this, too. It's that sun-and-Earth-thing where the Earth goes spinning about its own axis—we all learned that in elementary school. But once again, *knowing* the answer isn't the same as *understanding* it.

In reality, the reason the sun always sets in the west is complicated—incredibly, stupendously, ridiculously complicated—because every time you start talking about the spinning of the Earth you almost instantly find out you've got to first understand the basic laws of motion. That raises questions about how the sun stays there, hanging in the sky, without falling . . . *somewhere else.* Whoops, now that you bring that up, you realize we're talking about gravity here. Oh, boy, gravity—that's a tough one. Meanwhile, we're back to the setting sun: If you want to know why it always sets in the west, you really should make sure you understand how a compass works so you can be sure what you're calling west is, in fact, really west. And to understand the principles behind a compass, you again return to gravity—and you have to understand the nature of electricity and magnetism. And then, after you've read and studied and probed and think you have come to a full understanding of all these phenomena of the physical universe . . . you'll finally come to this chilling realization that you already had a long time ago anyway: *We're lost in space.*

The second most commonly asked question in physics: Why is physics so hard? Implied in this question, of course, are two other questions. First, this one:

WHAT IS PHYSICS?

It's a set of laws. Not only is it a set of laws, but physicists like to think that it is the smallest possible set of laws that can be found to describe the largest possible number of things. It's, like, getting down to real basics.

And when we say laws, we're talking the kind of laws that govern the universe—which means they're a lot different than the laws you lay down to rule the roost. And there's one simple reason for that: In your day-to-day existence, you make the rules. You might change the rules. And the rules are all based (we hope) on some sort of human logic or sense of fair play, tradition, or precedent. But around the universe, we don't know who made the rules. We don't know why the rules were made. We don't even know what function they serve. We just know the rules are the rules, and we're not allowed to break them. So ultimately, physics is a process of figuring out those rules—or refining the ones we've already got until we narrow things down to the simplest set of rules possible. Back to basics.

WHAT IS HARD?

Um, when you say physics is hard, what do you even mean by *hard*?

Believe it or not, this tough question gets to the very root of the study of physics.

Until the turn of the twentieth century, physicists would have told you that their idea of hard was the same as yours . . . and if you had any doubts as to whether something was hard, you could just go bang your knee against it, and if you got a bruise, that would be your answer.

In modern physics, *hard* is a little, um, harder.

It can be tremendously difficult for those who are not physicists to understand that even when you bruise your leg on a table, the table itself, at least on a subatomic level, is made up of a lot more space than it is stuff. But more, much more, on that cosmic stuff later.

Physics, on the other hand, which leaves no bruises, is one of the hardest things many people have ever encountered in their lives. So, just as in the English language, *hard* can mean hard as in "ouch!" or it can mean hard as in a real brain-stretcher.

Finally, the one question about physics nobody ever asks:

What if Physics Were Simple? Six Ways to Define Physics

When Nobel Prize–winning physicist Richard Feynman was asked by an artist friend why physics was so hard, Feynman was astonished. "Physics isn't hard," Feynman replied. "If it was hard, we wouldn't know so much about it."

Feynman's comments are instructive, but for those of us who are not scientific geniuses, physics is indeed hard. Here are several reasons:

- **The language of the laws of physics is mathematics.** Therefore, what can often be said simply in a few formulas and clearly understood by those with a math background may require thousands of words to explain adequately to those without a solid mathematical footing.
- **Physics . . . it's all about relationships.** "Sure it is," you say. "Should I pick up a copy of *Cosmopolitan* magazine and learn about physics there?" Actually, *Cosmo* wouldn't be a bad place to go if you wanted to do some additional reading. After all, discovering the principles and laws of physics is sort of like figuring out the rules of dating.

  What is dating, after all? It is a process of testing and experimentation that will lead to either commitment and marriage or a repeat of the same process with someone else. You cast a few lines out there, see someone you like, get to know him or her, you throw in a few surprises to see how he/she reacts, then you hang around long enough to determine whether this one will stand the test of time. If he/she's a keeper, you get married. If not, you throw him/her back into the pond, rebait your hook, and recast. Things work the same way in physics. Only instead of having significant others, physicists have theories.

  Actually, like most human relationships, physics carries a lot of baggage—a long line of unhappy and unsuccessful relationships with theories that turned out to be bad news. On the other hand, there are a few theories out there—like the theories of electricity and magnetism and universal gravitation—that have been around quite a while and have withstood the test of time.

Physics is also about the relationships that different elements of nature have with each other—for instance, the relationship between time and space, matter and energy, electricity and magnetism, force and speed. Yet the irony is that these relationships look just like the crazy, mixed-up ones most of us have on our hands; they're complicated, hard to understand, and subject to change (given our perceptions of them), and although we're convinced there's an iron-clad set of rules governing the way they work, we don't have a clue as to what those rules are.

- **Physics is the study of matter and energy and how they interact.** Look in any science textbook and it will tell you. But what are matter and energy? *Matter* refers to anything that takes up space, no matter how large or small it is. *Energy* refers to any force that can produce a change in matter.

As you can see, the definitions of *matter* and *energy* pretty much cover every single thing known to people (including our thoughts, which are waves in our brains that require energy to be generated). Physics, then, may include the study of anything you can see, hear, feel, touch, imagine, and think of. It is also the study of everything not covered in the previous sentence.

- **Physics is an attitude thing—we say we are "doing" physics when we get curious about some element of nature** (like why do toilet bowls in the Northern Hemisphere always flush counterclockwise but toilets in the Southern Hemisphere always flush clockwise?*) and then we begin asking questions about the behavior and do what ever it takes to find the answer.

- **Physics is the study of laws that either describe or govern nature; to test these laws, we use a series of better and better approximations that can be tested as a result of new**

---

* Okay, it's true—it doesn't always happen this way. For instance, it won't necessarily work if your toilet bowl isn't perfectly symmetrical or if, like those of us in California with low-flow toilets, you don't have enough water in your tank to cause the phenomenon to occur. Incidentally, if you are on the equator, the water in your toilet will run straight down. All of this is due to the Coriolis force, which is the influence of Earth's rotation on any moving body.

technologies. These new technologies provide a broader range of experimental information. Meanwhile, the development of new techniques may provide access to new information.

■ **Physics is like one of those Chinese boxes you open up and . . . inside is another.** For 2,000 years, physicists have persisted in opening box after box after box after box . . . and despite all they know they still haven't gotten to the last one.

Take Albert Einstein's theory of relativity, for example—what he was able to initially describe mathematically with just a few equations has been the subject of hundreds of books, many of which attempt to describe the same thing using words instead of math. Yet given all those books, how well do you think the average Joe on the street understands it?

Let's compare physics with another field of endeavor—say, accounting. When you are an accountant, the domain in which you work is one which was created by fellow humans; therefore, the systems you use to do your work are based on a system of logic with which you are familiar. The physicist, on the other hand, works in a domain that was created by . . . who? What? And because we don't know who created the natural world around us or the reason it was created, the physicist has to figure out not only why things are that way but also the logic of the system itself. If you saw the movie *Contact*, you will recall that scientists couldn't read the message that was sent from outer space until they figured out how to read it. Once they figured out that the message was sent in a three-dimensional format instead of what is traditional for us—a two-dimensional format—they were off and running. And, you'll recall, the message itself was easy for them to decipher; after all, it was written in the universal language of mathematics.

Physicists must often do their job without knowing if they are looking at the whole picture or even if they have all the facts.

One way to look at the difficulty of understanding physics is to think of it like one of those three-dimensional posters that appears to be only a wild psychedelic design until you stare at it long enough . . . and that's when the fish or the giraffe or castle or some other scene suddenly pops out at you. And once you see it, it is so obvious . . . it's hard to believe you couldn't see it just a moment before. Same thing with physics. You stare and stare and stare, get up to get a cup of coffee, and start staring again when—*poof!*—suddenly it all makes sense. Trust us, it will. Keep reading.

You Got a Problem?
How to Think in Physics

If you want to be a physicist, or just think like one, you've got to have a problem.

Maybe you want to know why apples fall from trees. Or how it's possible for the Energizer Bunny to keep going and going and going and going. Or why a football flies through the air with a perfect spiral when you throw it just right.

ALBERT EINSTEIN EXPLAINS HOW TO THINK ABOUT THINKING

What, precisely, is "thinking"? When, at the reception of sense-impressions, memory-pictures emerge, this is not yet "thinking." And when such pictures form series, each member of which calls forth another, this too is not yet "thinking." When, however, a certain picture turns up in many such series, then—precisely through such return—it becomes an ordering element for such series, in that it connects series which in themselves are unconnected. Such an element becomes an instrument, a concept. I think that the transition from free association or "dreaming" to thinking is characterized by the more or less dominating role which the "concept" plays in it. It is by no means necessary that a concept must be connected with a sensorily cognizable and reproducible sign (word); but when this is the case thinking becomes by means of that fact communicable.

"With what right—the reader will ask—does this man operate so carelessly and primitively with ideas in such a problematic realm without making even the least effort to prove anything?" My defense: all our thinking is of this nature of a free play with concepts; the justification for this play lies in the measure of survey over the experience of the senses which we are able to achieve with its aid. The concept of "truth" can not yet be applied to such a structure; to my thinking this concept can come in question only when a far-reaching agreement (convention) concerning the elements and rules of the game is already at hand.

—Albert Einstein

In physics, if you've got a question, then you've got a problem—and sometimes, one of the biggest problems in physics is making sure that you've asked the right question.

Sometimes, though, you can ask until you're blue in the face and you'll still never get a satisfactory answer.

Want an example? Take gravity. We know what it does. We know where to find it. We know how to use gravitational equations to predict how matter and energy will interact. But ask many physicists what it is and they'll merely shrug—for that is something that nobody knows. And physicists have been studying gravity for hundreds of years.

GALILEO'S SCIENTIFIC METHOD UNVEILED

In science, there's the right way to do things and then there's the real way. The right way is called "the scientific method." A guy named Galileo (1564–1642) is usually given credit for cooking the thing up, and on the whole, it's a process that really makes a lot of sense.

"I've got a theory about that . . ."

To use the scientific method, follow these steps:

1. Define a problem: You know, like "Why does an apple hit the ground after it falls from a tree?" or "How come I always get holes in my socks?"
2. Take a guess at the answer . . . maybe you always get holes in your socks because you never trim your nails?
3. Invent a way to test your guess (a guess is also known as a hypothesis). For instance, you might buy a dozen pairs of identical socks (and it's crucial that they be the exact same type of socks or you will have changed another variable). Wear one pair until you get a hole and take note of how long it took for the hole to appear. Trim your nails. Then wear a new pair of socks and see how long it takes you to wear a hole in them.
4. Compare your results with your hypothesis. If it takes longer to wear a hole in your socks after the pedicure, you've got a viable hypothesis.

5. Repeat the experiment over and over and over—enough times so that you feel comfortable going public with results.

6. If your socks don't last longer after you trim your nails, go back to step 2. (But don't throw in the towel too quickly . . . for one of the odd quirks of doing scientific research is that you never know if you've done enough trials until after you've done too many.)

The scientific method forces you to operate in a manner that is logical and well planned. It also helps you avoid one of the biggest pitfalls in scientific experimentation, which is to test too many variables at one time.

Whenever you perform a scientific experiment, it's important that you only change one variable in the experiment at a time. In the example above, let's say you skip the part where you wear a new pair of socks before trimming your nails. This would mean that you now have two variables at work in the experiment and no control with which to compare your results. The two variables are now your feet after your date with yourself to hack those nails back, and a new pair of socks after your handiwork is done. But what will you compare your results against? After all, if you run the experiment this way, you don't know how long a pair of socks lasted before you cut your nails, so you don't know the length of time they last after you've run your experiment.

Sometimes, narrowing your experiment down to a single variable is the hardest part of the job—because if you aren't careful, you'll end up testing for something other than that which you think you are. Case in point: You probably remember that in the 1970s someone did a test on house plants and discovered that if you talked to them nicely they would do better than if you yelled at them all the time.

At least that's the variable (tone of voice) the experimenters thought they were testing. Turns out they were wrong. After all the hue and cry had dissipated about the new, proven scientific fact that plants have feelings, later tests demonstrated that the methodology of these initial experiments had been wrong.

It seems out the plants weren't responding to the tone of the voice and message being directed at them—they were responding to the physical presence of the messenger. When we shout, we tend to make big, threatening movements toward the object of our anger. When we talk softly, we tend to make slow, fluid motions. Experimenters discovered that if one spoke softly and lovingly but moved toward the plants in a threatening manner, they would respond as if you had yelled and screamed at them. And vice versa.

So much for plant emotions . . . and flawed scientific evidence.

When you perform the sock-hole experiment as originally described, the only variable that changes are your nails. Therefore, if you repeat this experiment time and time again and there is consistently an increase in the life span of the socks, and then you repeat your experiment and verify your first result, then you can begin to attribute your findings to the fact that you clipped your nails.

As is often the case in life, the right way to get things done is not always the real way things get done.

For instance, as neat and orderly as Galileo's scientific method may be, the downside is the same: It is logical, neat, and orderly—but great scientific discoveries are often born of chaos and intuition and chance. Had Albert Einstein, a guy who was famous for being absentminded and eccentric, done all his research according to Galileo's step-by-step rules, he might never have discovered the theory of relativity.

HOW CARL SAGAN SEPARATED THE SCIENTIFIC TRUTH FROM THE BULL

Call it the scientific method or call it your "baloney detection kit," one absolutely central and key skill in the modern age is the ability to make some rough set of rules about how to go about winnowing out a little truth from the vast ocean of falsehood and confusion in which we live.

Here's how it works: Somebody comes up with a hypothesis which takes certain positions on a subject. If the hypothesis is worth talking about at all, others will test those positions to see if the hypothesis stands or falls. If the hypothesis falls, then you abandon it—you don't try to resuscitate it or find excuses for it. You just go on to something else.

That's because the scientific method is a delicate balance between an openness to all ideas, no matter how seemingly bizarre, and at the same time, the most rigorous skeptical scrutiny of all wisdom, ancient and modern. Different people have different predispositions in this regard—some people tend toward being open and some are more critical, and depending on which predisposition you have, it's important to remember to do the other half—and the key is to be good at both.

On the other hand, there are all sorts of studies that show that if we don't watch it, we are likely to believe anything that anybody in power tells us. You can see this in its darkest form in so-called cults—but you can also see it in all the people who believe what the leaders of this or that nation say, instead of applying the same standards of critical thinking that they would use if they were evaluating a beer or aspirin commercial on television . . . but there again, the reason that commercials work so well is that these skills are not well honed.

The ability to apply the scientific method is an acquired skill—nobody's born with it. Unfortunately, the scientific method is not taught today—you can go through twelve years of public school without ever coming upon it. My sense is that schools would rather not teach it because it makes kids ask embarrassing questions that teachers have trouble answering.

And that's the most unfortunate part of this whole story; it's very important to recognize that no one has ever been right all the time, including Einstein and Newton and the other great figures of science. For some reason, we mostly only teach the successes and not the failures, but the failures greatly outnumber the successes and it's important for people to understand that people believed that the Earth was flat and also the center of the universe and all the revered philosophical, scientific, secular, and religious figures of the time taught that and believed it passionately—but they were utterly and completely and dead wrong.

Ultimately, the scientific method has applications not just to science, where it's been responsible for many of the advances on which we depend in the modern world, but it is also enormously helpful when applied to politics, religion, and economics, and even things of everyday life—like buying a used car.

—Carl Sagan

HOW TO TALK IN PHYSICS

Although the language of physics often looks and sounds like standard English, the truth is that regular folks often mean one thing with a word when physicists usually mean something more narrowly defined. Let's start with some simple words.

ENGLISH-TO-PHYSICS DICTIONARY

WORD	ENGLISH DEFINITION	PHYSICS DEFINITION
Work	Where you go in the morning. And what you do after you get there (when you're not talking on the phone, surfing the net, grabbing a Coke, or out to lunch).	Just like down in Personnel, the word *work* in physics has its own formula: $$W = F \times d$$ Don't let the math scare you. W = work; F = force; d = distance. Work, therefore, is what happens when something (a force) causes something else (a body; see below) to move some distance. And here's the good news: According to this definition, if you just show up for work in the morning and lift your coffee cup to your lips, you have performed work and there's nothing Personnel can do about it if you don't do anything else for the rest of the day. Of course, if you do manual labor for a living—for instance, bus tables, load trucks, dig ditches—then the things you spend your days doing come closer to a physicist's definition of *work*.
Velocity	What you move with when you're late for work.	There are two parts of velocity: 1. Speed—which is the distance you go divided by the amount of time it took you to get there, or 60 miles per hour, for instance. 2. Direction—because, when you leave L.A. at 60 miles per hour you can either go north to San Francisco or south to

WORD	ENGLISH DEFINITION	PHYSICS DEFINITION
Velocity (cont.)		Baja. Same speed, different direction.
		Velocity is the combination of these two elements—speed and direction.
		Your first reaction to the concept of velocity might be "Who cares?" But what if you're out in the middle of a river on a raft in the middle of a storm? The current is moving at 10 miles per hour and there is only one place you can get to and dock safely. Suddenly, your velocity becomes very important because if you head directly toward the dock, you will surely miss it; the current will push you down river as you struggle to get across. But if you understand the concept of velocity, then you know to choose a direction of travel upstream of the dock, thus taking into account the combination of the river's speed (10 miles per hour) and the river's direction (downstream).
		With all of this in mind, here is the definition of velocity: it is a speed pointing in a particular direction—just like an arrow. An arrow has a length and a direction. A velocity has a speed and a direction.
		As you might guess, then, if you move in a straight line and never change direction, your speed and the magnitude of your velocity are the same.

WORD	ENGLISH DEFINITION	PHYSICS DEFINITION
Velocity (cont.)		Change direction, however, and suddenly your velocity is different. But the joke's on you because although your velocity is different, the speed and magnitude of your velocity remain the same.
		Confused? Try this on for size: If you twirl a ball on a string over your head, it will move at a constant speed but its velocity will be ever changing because the direction the ball is traveling is continually changing—even if the magnitude of velocity (a.k.a. speed) remains constant.
Acceleration	What you wish you had more of when you're late for work.	Acceleration is not a difficult concept . . . in fact, there's a gadget in your car that makes acceleration easy to understand—it's called an accelerator.
		Press on it and you go faster. Your speed has changed. Take your foot off it and your speed decreases. This seems easy to understand when we are standing still and step on the accelerator—especially if we're in a Porsche 911, which goes from 0 to 60 in about 5.2 seconds. What is it that causes our hair to fly back behind us and pulls the skin of our faces toward the back of our heads?
		Acceleration, of course. You stomp your foot down on the accelerator and man, this baby goes! Acceleration becomes a little more confusing when you consider what happens if you're

WORD	ENGLISH DEFINITION	PHYSICS DEFINITION
Acceleration (cont.)		on the highway in your 911 and there is a highway patrol officer in the next lane. Wisely, you set your cruise control for 65 miles per hour. Since you are going a constant speed—65 miles per hour—there's no acceleration. When it comes to acceleration, from a physicist's point of view, a constant speed is a constant speed, whether that speed is 0 or 65 miles per hour—in both cases, the acceleration is nill.
Pressure	This is what your boss puts on you when he wants you to do more work.	Actually, forget your boss—that ain't pressure. Pressure's what's going on when that mail cart rolls over your foot. Pressure is the force of an object divided by the surface area with which it is making contact. When a cat lies on your stomach, your stomach generally doesn't hurt, but it can hurt if the cat *walks* on you. The weight of the cat does not change—only the surface area with which it makes contact. Since the surface area is decreasing, the pressure must be increasing. Pressure's also what's going on in a bottle of Coke. Its pressure comes from the fact that all the little gas molecules are banging against the sides of the bottle—pushing the bottle outward.
Force	What your boss will resort to if pressure doesn't work.	Let's say you are rolling your mail cart down the hallway at work when suddenly the UPS delivery person runs into you with a dolly. The mail cart

WORD	ENGLISH DEFINITION	PHYSICS DEFINITION
Force (cont.)		immediately changes direction and rolls over your foot.
		When the dolly hit your cart, it applied a force—it gave your mail cart a push and caused it to move at a different speed and in a different direction than the speed and direction it was moving before the delivery-person came along.
		But note: if the UPS delivery-person hit your cart when she and you and your cart were going in the same direction, that would still be force because your cart would end up going faster in the direction it was already moving than it was going before.
		Anytime the speed and/or direction of a body changes as a result of an influence from another body, that influence is considered force.
Mass	Where you go on Sunday. Or the number of people at your favorite bar on Saturday night.	Mass refers to how much matter is in a body. This may sound funny, so to clarify a bit, let's imagine you are the mail clerk at work and your job is to push a cart around and distribute the mail.
		On most days, your mail cart is full of letters. Full to the brim. Once in a while, though, some-one stacks it full of hardcover books, and you're the lucky soul who has to deliver them. Same cart. Filled to the same level with stuff. Yet when the cart is full of books, it's way harder to push.

WORD	ENGLISH DEFINITION	PHYSICS DEFINITION
Mass (cont.)		Why? Because the books have more mass, more matter, in them than ordinary letters do.
		Another way to define mass is as the tendency of a body to resist force.
		It is harder for you to push a cart of books than it is to push a cart of letters because when you push, you are exerting a force on the cart, yet the books have more mass.
Weight	That about which you are in a constant state of denial.	Now you're in for a real surprise: your weight to a physicist has nothing to do with how you look in a bikini or a pair of shorts. Instead—are you ready for this?—weight is a force. And this might be even more surprising to you: Weight is not just any kind of force; it is a very special kind of force called gravitational force.
		So the good news is that where you tip the scale has nothing to do with what you ate for lunch and everything to do with how hard the gravity of the Earth is pulling down on you. Which of course has everything to do with what you ate for lunch (so that good news goes right out the window, after all).
		And if you don't like that, go tell it to Jenny Craig.
Body	The object of your denial.	Anything with mass (see above)—in other words, you again. And everything else around you.

PHYSICS ACROSS THREE TIME ZONES

If you wanted to make a timeline of the history of physics, it would look something like this:

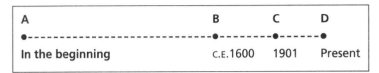

A			B	C	D
●-----------------------------●---------●------●					
In the beginning			C.E.1600	1901	Present

PHYSICS TIMELINE EXPLAINED		
A → B	B → C	C → D
Physics is the mother of all sciences. In the days of yore, physics was part of a field called natural philosophy and it actually encompassed all of the sciences, from the life sciences (like biology, physiology, and zoology) to the physical sciences (like chemistry, astronomy, and geology).	**What a classic!** It all began with Galileo, who wasn't so much interested in studying the stuff itself as he was interested in studying how and why that stuff moves.	**Jet-lag physics.** In 1900, Max Planck introduced his quantum theory (which inquires into particles that are very small). In 1905, Einstein introduced his theory of special relativity (which inquires into phenomena that are very fast). Thanks to these two guys, things just haven't been the same around here since. Spend enough time studying these two theories and you're bound to feel like you've crossed too many time zones.
Three big questions. Early physicists had three things on their minds: ■ What is this stuff lying all over the place? And what is this stuff made of? ■ Why do things down here on Earth move the way they do?	**One big question:** "Hey, you don't think there could be any relationship between the stuff down here and the way it moves around and the stuff up there in the sky and the way it moves around, do you?"	**A million different questions.** Beginning with: "Atoms, huh? Well, if what you're looking at is an atom, then what do you think is inside of it?"

A → B	B → C	C → D
▪ What's up with the sun and moon? Why do they always move around like that?		

TWO THOUSAND YEARS OF WRONG THINKING IN PHYSICS

For the first two millennia, physics was a slow-moving science. Which is no wonder, since early physicists spent their first 2,000 years banging their heads against the wall (which felt hard to them and, in fact, left bruises). They were trying to prove that a lot of things were right—when in fact they were wrong.

THE WAY THINGS WEREN'T		
How (Not) to Build a Universe	"Did You See Where I Left My *Apeiron*?"	Introducing the World's First Big-Screen TV
Everyone knows that if you're going to build, the trick is to hire the right contractor and make sure he uses the right materials: ▪ Obviously, Thales of Miletus (c. 625–545 B.C.E.) should have talked to the guys down at Home Depot before he purchased his supplies for the construction of his universe. He thought that everything in the universe was made of water. ▪ Thales' neighbor, Anaximander (c. 610–545 B.C.E.), had a different idea; he came	Funny how life works. Thales and Anaximander only had one kind of stuff to keep track of and Empedocles had only four, yet all three of them kept losing their stuff because they didn't understand how and why it kept moving around.	Imagine a world without all your favorite TV shows. No wonder early physicists spent so much of their time staring off into space. Naturally, just as *TV Guide* and *Entertainment Weekly* and *People* magazine make it easier for all of us to talk about what we watched last night, physics provided people way back when with a topic around which they could talk about what they watched last night . . . which was the movement of the moon and the stars. Of course, if you want to hear about it from

How (Not) to Build a Universe	"Did You See Where I Left My *Apeiron*?"	Introducing the World's First Big-Screen TV
to the conclusion that the one essential ingredient you needed if you were going to build a universe was something he called *apeiron*—but he never said what *apeiron* was. For a literal translation from Greek, try "the boundless" or "the indefinite," an answer that, we are sure, must have gotten Anaximander a really bad grade on his final exam. ■ While at a barbecue at the beach, a guy named Empedocles of Agrigentum suddenly had a thought: "Wouldn't it be cool if every day could be like this?" Thus was born the theory that everything in the universe is made of four elements: fire, air, water, and earth.		someone who's been to the moon and back, read *Flying to the Moon and Other Strange Places* by Apollo 11 astronaut Michael Collins. Just want to read about it from someone who got close? See Jim Lovell's book *Lost Moon: The Perilous Voyage of Apollo 13*.

ARISTARCHUS AND ARISTOTLE

When it came to all that movement up there in the sky, one guy had a different idea. His name was Aristarchus; he lived around 310 to 230 B.C.E. and had this crazy idea: "What if the Earth wasn't at the center of the universe . . . what if all the planets revolved around the sun instead?" What a guy, this Aristarchus! The sun at the center of the universe? What a nut!

Instead of finding a receptive audience, he was persecuted by the

Stoic philosopher Cleanthes and ignored by his fellow physicists. In fact, if other scientists learned anything from Aristarchus, it was to keep their own mouths shut. But because Aristotle lived earlier (384–322 B.C.E.), he hadn't learned that lesson—he gave civilization plenty of wrong ideas.

MISSING THE BOAT WITH ARISTOTLE		
Aristotle's Theory of Motion	Aristotle's Falling Body Hypothesis	Aristotle Explains the Movement of the Heavens
As Aristotle saw it, every body had a natural place in the universe and bodies stayed in their natural places unless some external force caused them to move. Aristotle called it violent motion when a force caused a body to move from its natural place. Meanwhile, Aristotle said, if you remove the force, the body anxiously returns to its natural place—what Aristotle called natural motion.	According to Aristotle's theory of motion, the speed at which a body returned to its natural place after being forced to move was dependent on its weight. One manifestation of this, he told us, was that if you were to drop a really heavy object and a really light object at the same time, the heavy object would reach the Earth much faster than the light object.	Aristotle applied his theory of motion only to objects on Earth, of course. When it came to the heavenly bodies, he claimed that because they moved in perfect spheres without beginning or end, they were demonstrating another form of natural motion.

Aristotle was wrongest for the longest when it comes to theories about motion here on the ground or to ideas about the sun and the moon. But the guy who wins the prize for the longest-running case of wrong thinking when it comes to theories about the movement of the planets was a Greek chap by the name of Ptolemy, who died around C.E. 178.

PTOLEMY WHAT YOU THINK

Ptolemy had to stack the deck of his universe with orbit after orbit after orbit to explain such irregularities of the nighttime sky as

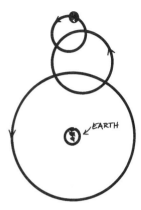

retrograde* motion. Draw a diagram of Ptolemy's theory about the Earth and the sun and the planets and you get something that resembles an early celestial Spirograph.

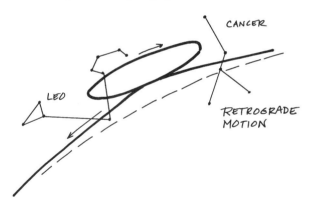

Wrong or not, Ptolemy had a thousand years' worth of followers, including Theon of Alexandria (c. C.E. 400); his daughter Hypatia; and the Arabs Harun ar-Rashid (c. 800), Caliph al-Mamun (c. 829), Albumazar (805–885), Abd-ar-rahman al-Sūf (903–986), Ibn Junis (c. 950–1008), Nasir ud-din (1201–1274), and Ulugh Beg (1394–1449).

---

* Have you ever noticed that a planet like Mars will move eastward across the sky for some number of months and then, without reason, slow down and go back in the other direction? This is what physicists and astronomers call retrograde motion and it all makes perfect sense when you put the sun at the center of the universe. But when you put the Earth at the center of things, what you get is a horrible mess, as this brief summary of celestial history shows.

The Moors liked what they saw and took the Arabs' brand of Ptolemaic astronomy with them to Spain, where it also flourished. It then made a leap to Germany, where astronomer George Peurbach (1423–1461) took the fallible theory to heart.

ARISTARCHUS GETS VINDICATED AND PTOLEMY GETS A KICK IN THE PBUTT

Some 1,750 years after Aristarchus took it on the chin (and about 1,300 after Ptolemy threw his wrench into the works), a guy named Copernicus came along and made the same claim: "Hey, the Earth isn't at the center of the solar system . . . the sun is!" Funny thing was, he was right.

LET THE CLASSICAL PHYSICS REVOLUTION BEGIN: FROM GALILEO 2 NEWTON

Galileo wasn't afraid to stand up and be counted. He knew Copernicus was right, and he said so. In return, Galileo got no more respect than Aristarchus did—he was dissed by the church and put under house arrest for the last 10 years of his life. But this was just one event in a long career of fun and trouble in the pursuit of good science.

GALILEO DEVELOPS THE FIRST ASTRONOMICAL TELESCOPE

Although it is easy for us to take the miracle of magnification for granted, the significance of the first telescope for looking at the planets and the stars shouldn't be discounted. It was the first time in the history of humankind that the heavens could be seen more closely than with the naked eye alone. On the basis of what could be seen through his telescope, Galileo claimed that he had proved Copernicus to be right—the sun was at the center of the universe, and the Earth and all the other planets were moving and revolving around it.

SCIENCE THROWS DOWN THE GAUNTLET

It's easy to see why the church would be threatened by these theories. The universe is a really big place, and as long as the way it worked remained a mystery, then it was easy for the church to claim that it was the hand of God that kept it in order. As scientists got better at proving that the universe is actually a mechanical system that runs like a clock, the more the church saw the potential for its power to slip away.

GALILEO TAKES ON THE CHURCH . . .
AND PUBLIC OPINION

Here's how the public reacted to Galileo's theory: "The Earth is moving? C'mon, Mac, give us a break! If the Earth is moving, how come we can't feel it move? How come when I jump up in the air, I land in the same place I jumped from—instead of a few feet or a few inches behind my starting point? Huh?" And the church was even less receptive.

To answer his critics, Galileo proposed to climb up the mast of a moving ship at sea and drop a ball. His mission was to draw an analogy between the moving ship and his hypothetically moving earth. "If," said Galileo, "a moving Earth would cause a falling object to fall a small distance away from where it was dropped, then it should also reason that if I climb to the top of the mast of a moving ship and drop a ball, the ball should fall some distance behind the base of the mast. However, if the ship is moving and I drop that ball and it falls directly to the base of the mast, then this would be analogous to why you can drop a ball on a moving Earth and have it fall to your feet or how you could jump up and land in the same spot on a moving Earth."

Galileo boarded a ship. Sailed with it to sea. Climbed up the mast while the ship was moving and dropped his ball. And it landed just where he said it would . . . at the base of the mast.

GALILEO TAKES ON ARISTOTLE

Sometimes in science, the only way you can get things done is to destroy the work of someone who came before you. For 2,000 years, Aristotle's assertions about the nature of motion were accepted blindly. But why test them? Aristotle, after all, was the greatest teacher of all time, so what sane person would question him?

Galileo realized that the old falling-ball-from-the-mast stunt wasn't going to be enough to convince people that the Earth was moving. To really clinch the deal, he knew he was also going to have to show the people that Aristotle's notion of motion was incorrect. After all, people had believed that Aristotle was right about the nature of motion for over 2,000 years, so Galileo knew no one would believe anything different until Aristotle was proven wrong. And here was the real challenge: According to Aristotle's theory, the only way the Earth would ever move from its natural place was if there was another force hanging around that was strong enough to push it . . . and what force could Galileo possibly point to that would be large enough to do that?

"WHAT IF PHYSICS WERE DIFFERENT?"

Consider the plight of poor physicists. Their job is to think the unthinkable and to imagine the unimaginable—and the greatest tool at their disposal is the question "What if . . . ?"

Then they set out to find the answer.

Imagine you were Galileo, trying to explain the laws of physics as they apply to a falling rock or ball.

Hmmmmm, you might think to yourself. *What I really need here is a law that explains acceleration.*

Acceleration?

An easy enough concept for most of us to grasp: we merely have to consider a plane as it takes off or jump in a car and stomp on the gas and the concept of acceleration becomes instantly clear. Acceleration is a part of our everyday lives.

The same was not true for Galileo. The only time he could observe acceleration was when he dropped something from a high tower or cliff, or very briefly as he watched a horse and rider. And even then, his opportunity for observation would last only a few seconds. Oh, and yes, it's also important to remember that he didn't have a stopwatch with which to record the time it took balls to fall from various heights (although it is rumored that he would sometimes time things by taking his own pulse). Nor did he have a video camera to allow him to tape-record a ball's fall and then replay it over and over again at various speeds.

Instead, he had to use his mind to imagine what would happen if he could slow down the ball's fall. And using the power of abstract thinking, he realized that he really could slow the fall of the ball in reality by rolling it down an inclined plane. (But was that the right approach? After all, Aristotle slowed motion by dropping objects into a viscous fluid—where friction played a dominant role.) But back to Galileo: To support his argument that all bodies fall at the same speed in a medium without resistance (a.k.a. a vacuum, which no one but Galileo believed could exist) he argued that if performing experiments in a medium of greater resistance, such as water, reduced the speed of a falling ball, then one could make assumptions about the speed of a falling ball in a medium of no resistance.

Today's physicists face the same hurdles as Galileo. Their job is to describe and explain parts of the universe that previously have gone unnoticed or unseen or continue to be inexplicable. For them, too, picking the right question is every bit as important as finding the right answer.

GALILEO STARTS SMALL

Instead of attacking the whole of Aristotle's theory, Galileo set out to discredit just one small but crucial part of it—Aristotle's notion that the heavier an object is, the faster it will fall to the ground. To do this, Galileo took both a heavy and a light object to the top of the Leaning Tower of Pisa and dropped them at the same time. Although the heavier of the two did fall slightly faster, the difference in speed was tiny and in no way proportional to the difference in weight between the two objects. (Pretty good story, huh? The funny thing is, no one knows if it's true or not.)

For conversation's sake, let's say Galileo dropped two stones from the tower. Furthermore, let's say one stone was 10 times heavier than the other. If Aristotle's theory was correct, the small stone should have taken longer to hit the ground than the larger stone, with the difference being proportional to the difference between their weights. Or from the other perspective, at the point where the large stone hit the ground from the 180-foot-high tower above, the small stone should have fallen only 18 feet. But this is not what happened. Instead, the large stone hit the ground just an instant before the small stone and Galileo came to a logical (and, it so happens, correct) conclusion: The difference in time had nothing to do with weight.

GALILEO'S NEXT PROBLEM . . .

Galileo didn't believe in Aristotle's notion that a body—any body—will remain at rest in its natural place unless a force comes along and causes it to move. The problem with this theory, said Galileo, is that for some bodies, the natural state is to be in motion.

And here's the kicker: Galileo topped off his theory of motion by declaring that a body at rest will stay at rest unless something comes along to disturb it or a body in motion will continue to move at the same speed and in the same direction until something comes along to cause it to speed up, slow down, change direction, or stop. Those of you who were paying attention in your high school science classes will instantly recognize this as the definition for Galileo's law of inertia.

Revolutionary thinking, but how could Galileo prove his theory was true? He knew he couldn't prove it by dropping more balls off the tower . . . after all, the balls hit the ground so fast it was impossible for him to make any observations about the fall. Somehow, he would have to find a way to make the balls fall farther . . . or better yet, slower. Sounds crazy, for Galileo had already said earlier that all objects fall to the ground at the same speed—but somehow, he had to slow that speed down. His solution: Instead of dropping balls, he would roll them down inclined planes.

By doing so, here's what he found: (1) a ball rolling down an inclined plane will pick up speed, and (2) a ball rolling up an inclined plane will lose speed.

On the basis of these two findings, Galileo came up with his first conclusion: If a ball rolling down an inclined plane picks up speed and a ball rolling up an inclined plane loses speed, then a ball rolling across a perfectly level plane shouldn't pick up speed or lose it—it should just keep rolling across that flat plane in the same direction and at the same speed forever. Of course, Galileo could never demonstrate this in reality because friction always caused the ball to slow down and eventually stop. What Galileo could do was demonstrate that the more friction the ball encountered on the surface of the flat plane, the slower the ball would go. The opposite was also true: When Galileo rolled a ball across an extremely smooth, flat surface, it rolled even farther. From this, Galileo concluded that if he could roll a ball across a plane that was perfectly flat in a frictionless environment, the ball would roll on forever unless some force caused it to change the speed of its motion.

Next Galileo positioned two inclined planes so they faced one another at equal angles, almost forming a U, and he let a ball roll down one plane and up the other. By doing this, he discovered that the ball would roll up the second plane almost to the same height from which Galileo had released it on the first plane. This observation led Galileo to another theory: If he could do this experiment in an environment where there was no friction between the ball and the plane, the ball would roll up the second inclined plane to exactly the height it had been released on the first plane.

To support this claim about the effects of friction, he repeated the experiment using smoother and smoother planes—and each time he used a smoother plane, the ball would come closer to reaching the level on the second plane at which it had been released on the first.

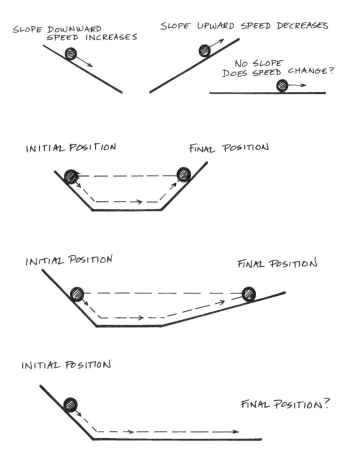

Galileo repeated this experiment many times. Sometimes, he would increase or decrease the slope of the planes or vary the horizontal distance between the two planes. In every case, the ball would roll down the first plane and up the second plane, stopping just short of the height from which it had been released.

This led him to his next conclusion: If a ball always rolls up the second inclined plane to just about the same height as it was released from the first plane—regardless of how much flat, horizontal distance is between the two planes—then if you roll a ball down one plane and there is no inclined plane for it to roll up, it will roll on forever without slowing or stopping. (Well, at least it should work if you could repeat the experiment in a frictionless environment.)

Let's call Galileo's ball "planet Earth." Somehow, Galileo had to relate the movement of those balls to the movement of the Earth. Here's how he did it: Since the Earth was traveling in a straight line, he said, then its movement was just like a ball rolling on forever on a frictionless surface—it would and could continue to roll forever. (Wait! Hold on a minute! The Earth . . . traveling in a straight line? Well, yes, however briefly. Be patient and hold on tight . . . we promise you'll understand in a few minutes.)

To keep rolling with the flow and find out more about inclined planes, see Michael S. Dahl's book *Inclined Planes*.

GALILEO AS SUPERMAN

So far, Galileo had used the ball-and-the-mast trick to suggest that the Earth could be moving without his fellow humans' being aware of it. Next, he tried to prove that if the Earth (or anything else) was moving, it would continue to move in a straight line without ever stopping. But how do you combine the two rules? According to Galileo, when you drop the ball from the mast of a moving ship, it falls straight down to the base of the mast whether the Earth is moving or not. Meanwhile, Galileo said, something else must happen when you drop the ball from the mast of a moving ship onto a moving Earth. It must also share the same straight-line motion as the Earth until something causes it to stop.

Ouch! It looks like Galileo had a problem. How can the ball fall two different ways at once?

To solve it, he proposed the principle of superposition, which stated that if an object is subjected to two different forces, the object will respond to each without changing its response to either. The best example of this is the firing of a bullet from a gun. Obviously, the bullet has been subjected to a force that propels it in the direction the gun is pointed. It is also true that the bullet will be subjected to the same force that pulled Galileo's balls down to the bottom of the Leaning Tower and to the bottom of his inclined planes (but we won't call that force gravity here, for *gravity* wasn't a term Galileo was familiar with). If the principle of superposition holds, then if you drop a bullet from shoulder height out of one hand and fire a gun held at shoulder height and pointed parallel to the ground with the other hand, the two bullets should hit the ground at the same time.

If the bullet example above is true (which it is), then you should be able to jump in the air and land in the same spot regardless of the movement of the Earth (which you can). But before we move on, let's make a picky point: If you shot the gun on a boat at the exact moment at which the boat veered sharply off to the left, then all bets would be off. The bullet you dropped from shoulder height would fall as before, but it would continue to fly off in the direction you had been heading before the boat veered.

BUT GALILEO DOESN'T COMPLETE THE CIRCLE

Granted, Galileo accomplished a lot, but he was never able to finish the job. Although it was true that he proved Aristotle's theory about natural motion was wrong and provided evidence that his own theories of inertia and superposition were right, he was never able to prove that the Earth was moving around the sun . . . only that it *could* happen.

This seems like a good time to make a point about the word *proof,* particularly when it is applied in the field of physics. To do this, we'll compare the concept of *proof* in physics with the concept of *proof* somewhere like in a presidential sex scandal, where the parties involved always assert their innocence and claim they can "prove" it. In physics, on the other hand, no one ever expects anyone to prove anything. The best we ever expect to be able to do is to "provide evidence" that what we are saying/thinking/asserting/hypothesizing/speculating is true. And in science, we are never too surprised when someone "proves"—or should we say when someone "provides evidence"—that we are wrong. After all, that's the scientific way. Compare that with the surprised look on a recent president's face when someone "provided evidence" (actually, on the basis of the info in the Ken Starr report, we're willing to call it proof) that he, in fact, had done exactly as everyone believed.

KEPLER IS LOYAL TO THE CHURCH BUT SUPPORTS GALILEO, TOO

It was the love of God—backed up by a huge collection of supporting data—that caused Johannes Kepler (1571–1630) to believe that the sun, not the Earth, was at the center of the universe. Why wouldn't God put the sun at the center of the universe, Kepler wondered to himself . . . after all, what is the Earth when compared to the sun, which is the giver of heat and warmth and light?

On the other hand, where Kepler did break from Copernicus was in his theory about the shape of each planet's orbit. Instead of assuming that the orbits would be in the shape of Aristotle's perfectly formed circle, he studied the data on the positions of the planets that were available to him (much of which he inherited—sort of—on the death of his boss, the eccentric Tycho Brahe, in 1641) and came to the conclusion that the shape of a planet's orbit is an ellipse. He was right.

Kepler's ellipse had two fixed points inside called foci (one such point is called a focus). It is the nature of an ellipse that if the inside perimeter were lined with mirrors, you could shine a light from one focus to any point on the ellipse and the light would bounce from

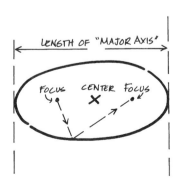

that point to the other focus. Cleverly, then, Kepler placed the sun on one of the foci, with some fascinating geometric and mathematical repercussions as a result.

The next thing Kepler did was build on Galileo's notion of inertia. Although Galileo suggested that an object that is moving will continue to move and a stationary object will remain stationary unless a force is applied in either case, Kepler redefined the term *force*. Force, he said, is not something that one body inflicts on another. Instead, it is an equal and mutual repulsion or attraction that exists between two bodies.

The combination of Kepler's ellipses, foci, and concept of mutual attraction between objects spelled one word: gravity. Too bad Kepler didn't realize it the time. If he had, little kids all over the world today would make faces when their parents serve them Fig Keplers after school.

DESCARTES ADDS HIS TWO CENTS AND FINALLY GETS AHEAD OF THE HORSE

You might think of René Descartes (1596–1650) as the inventor of the action movie, because even though he found what Galileo and Kepler were doing to be interesting, he was much more interested in figuring out what happens to things when they run into each other. Here's the result of all his figuring: When two objects collide, momentum from one may be transferred to the other, but no momentum will be lost as a result of the collision. Descartes called this "the law of conservation of momentum," or "the conservation law." If you apply this law to a game of pool, the speed at which the cue ball is rolling when it strikes another ball will equal the combined speed of both balls immediately after the two collide, provided they have the same mass. Another way of looking at it is to go back to our notion of velocity—once you understand that, you've also mastered all the hard parts of momentum. All you need to do is add some mass: Momentum is speed times mass in a particular direction. (Can you see the velocity in there?) Turns out that force is the rate of change of momentum—thus, if you had the choice of throwing yourself in front of a train or a VW Beetle that were both moving at the same slow speed, the Beetle would be the smart choice because the train would have way more mass . . . and thus, it also has more momentum with which to crush you. But if the train were moving only an inch per hour and the Beetle were moving at 200 miles per hour . . . well, gee, then you might want to reconsider your answer.

NEWTON WINS THE GLORY

Isaac (later to be Sir Isaac) Newton was born the same year Galileo died. And at the age of 23, he took all of Galileo's laws, mixed in Kepler's ellipses and the conservation rule of Descartes and Christian Huygens (1629–1695), and used them to change the face of physics forever by developing his own three laws of motion and the universal law of gravitation. This allowed him to do what Galileo could not: He provided evidence that the sun was at the center of the solar system and the Earth (and the other planets) were moving around it.

"SOLAR" SYSTEM? WHAT A CONCEPT!

If you want to prove that the sun is at the center of it all, you've first got to prove that you understand the laws that govern motion down here on Earth.

Law #1

The first law is simply a restatement of Galileo's law of inertia. According to Newton, every body continues in its state of rest or its state of uniform motion in a straight line unless it is compelled to change that state by forces impressed on it. In other words, a body continues to do whatever it was doing until something comes along to force it to do otherwise. Resting bodies rest and moving bodies continue to move uniformly until a force is applied to alter that state.

Law #3

For simplicity's sake, let's take Newton's laws slightly out of order. In his third law, Newton restates Kepler's idea of a mutual attraction between objects, but this is where many people get confused. After all, anyone who has ever used a hammer to drive in a nail knows that you are using the inertia (or momentum) of the hammer to apply force to the nail. But, Kepler would say, "That's only half the story," and Newton would agree. As they see it, the hammer and nail are applying equal amounts of force on each other.

Think this sounds crazy? Then look at it this way: Before a hammer can drive in a nail, it must accelerate—this is you raising your arm and then bringing your arm and the hammer downward toward the nail. When the hammer strikes the nail, it is exerting a force on it. But wait a minute—what about the law of inertia? If the nail didn't get in the way of the hammer, the hammer would just keep moving in the same direction forever. Therefore, the nail must be exerting a force on the hammer as well, causing it to decelerate.

In Newton's own words, his third law works like this: Whenever one body exerts a force on a second body, the second body exerts an equal and opposite force on the first. In today's language, we say, "For every action there is always an equal and opposite reaction."

Suddenly, the idea that physics might be about relationships doesn't seem so preposterous.

Law #2

Here, Newton adds the conservation rule to the mix. He said that the acceleration of a body is directly proportional to the net force acting on the body and inversely proportional to the mass of the body and is in the direction of the net force.

More relationship stuff. Don't worry if it doesn't make sense to you. We'll talk more about his second law in a minute.

NEWTON STEALS MORE IDEAS FROM GALILEO

What Newton is really doing here is talking about speed, acceleration, and velocity in a whole new way. Although Galileo had defined the terms during the course of his inclined-plane and falling-balls experiments, it was Newton who was able to apply the concepts to the workings of the heavenly spheres.

HOW TO THINK LIKE ISAAC NEWTON

When Newton said, "The acceleration of a body is directly proportional to the net force acting on the body . . ." what he meant was this: Suppose you apply a certain force to an object, and it accelerates with some acceleration. If you were to double the force on that object in the same direction in which it is traveling, the acceleration would double; if you were to decrease the force by, say, a factor of three in that same direction, the acceleration would drop to one third the original value. Note that this does not say anything about the speed of the object itself. Newton's second law does not deal with speed—it deals with changes in the speed and the direction.

When Newton said, ". . . and inversely proportional to the mass of the body and is in the direction of the net force," what he means is that the actual effect (or acceleration) caused by X pounds of force inflicted on our object will vary depending on our object's mass—the greater the object's mass, the less the effect—and vice versa.

In short, how much a force changes the motion of an object depends on both the size of the force and the mass (size) of the object.

THE WORLD ACCORDING TO NEWTON

Now, it's time for Isaac Newton to do what Galileo couldn't . . . to demonstrate not only that certain phenomena do occur but also to explain why.

PHENOMENON, PHENOMENA		
Mass Matters	**Friction Counts!**	**Gravity, the Missing Link**
When Galileo dropped those stones from the Leaning Tower, he showed that objects of different weights fall at about the same rate of speed. Newton used his second law to explain it. When Galileo dropped the stones, the heavier one was attracted to the ground with 10 times as much force ("The acceleration of a body is directly proportional to	Like a car dealer advertising on TV, we have repeatedly thrown in disclaimers when taking about the results of Galileo and Newton's experiments. That's because we can replicate these experiments only in a vacuum or frictionless environment. Otherwise, actual performance may vary.	Suddenly, all of Galileo's problems with friction begin to make sense. What Newton knew that Galileo didn't is all contained in his second law: The acceleration of a body is directly proportional to the net force acting on the body and inversely proportional to the mass of the body and is in the direction of the net force. Although it

Mass Matters	Friction Counts!	Gravity, the Missing Link

the net force acting on the body . . ."), but it also has 10 times greater mass, so it resists that acceleration 10 times harder (". . . and inversely proportional to the mass of the body and is in the direction of the net force"). Therefore, no matter what the difference is in the weight or mass of two objects, they will always fall at about the same rate of speed because when any object falls, its mass and resistance cancel each other out and the net force is the same. (Newton's law works best in a vacuum or frictionless environment, of course, but the principle still holds true in the real world.)

Let's go back to our two stones again. Let's say one weighs 10 pounds and the other weighs 100 pounds. What happens when we drop them both from the Tower of Pisa? They both hit the ground at about the same time. As we said before, if Aristotle were right, the 10-pound weight would have fallen only 10 percent of the distance to the ground at the moment when the

Galileo was the first to speculate on what might happen if he could remove friction. And he also speculated that the results of his experiments were correct—in a world without friction! What he couldn't do was explain why the friction was there.

Once again, Newton stepped in to save the day, demonstrating that friction is not the manifestation of gravitational force. Instead, he said, air friction is an independent force that acts in the opposite direction of a body's motion. So in Galileo's experiment, the force of gravity is pointing downward but the force of air friction is pointing up. That means the net force exerted on the balls Galileo used in his experiment is a little bit less than the pure force of gravity.

never occurred to Galileo that the Earth itself could be pulling on those balls as he rolled them down his inclined planes, it was very clear to Newton. Gravity also made it possible for Newton to explain why the sun, not the Earth, is at the center of the universe. (Don't worry if you don't see how—there'll be more on this later.)

Mass Matters	Friction Counts!	Gravity, the Missing Link
100-pound weight struck the Earth. What accounts for the small difference in time between the time the 100-pound weight and the 10-pound weight hit the ground is air resistance. This also explains why a 10-pound feather and a 10-pound cannonball will hit the ground at different times even though they have equal weight . . . given the mass of the feather, it faces much more air resistance because it occupies a larger volume— causing it to fall more slowly. Or, in a more practical example, the mass of a parachutist doesn't change after his chute opens . . . but the speed at which he falls sure does!		

GRAVITY . . . AS YOU'VE NEVER KNOWN HER

Mostly, we think of gravity only when we've just fallen victim to her— right after we've taken a header down a flight of stairs, dropped a jar of applesauce onto the supermarket floor, or been beaned on the head by a chunk of metal from a satellite somewhere.

And that's about as much as most people know about gravity. But the surprising thing is that gravity plays a huge role in our everyday lives (in addition to when we have accidents) and most people are completely unaware of it.

So what is gravity?

Newton really put his finger on it with his third law: Whenever one body exerts a force on a second body, the second body exerts an equal and opposite force on the first. To make it a little digestible, let's break it down into little pieces:

- A body, as we know, is simply some collection of matter. Could be your foot, could be your car, could be a tree in the woods. Could also be a single molecule of water. If it's matter, it's a body.

- Talking about force, on the other hand, is a little trickier. The easiest way to define force is to say it is an event that occurs when one collection of matter pushes or pulls another collection of matter. This could be an intentional push or pull, sort of like two kids locked in a schoolyard fight, or it could be accidental—my car bumps your car while I'm trying to parallel park. Regardless of how it happens, anytime one collection of matter pushes or pulls another, you've got force. Granted, this is a simple definition, one that would have made a guy like Galileo very happy. But Newton was a little more sophisticated than Galileo, so a better definition was required. As Newton saw it, force was not really a push or a pull; what it really was was an interaction between two collections of matter that resulted in a push or pull. Another way to describe this interaction between two objects: a relationship.

Next time someone tries to sue you because he or she tripped over a crack in your sidewalk, you might just show up in court with a copy of this book hidden away in your hip pocket. After the plaintiff has stated his or her case against you, tell the judge that you plan to countersue based on the testimony of your expert witness, Isaac Newton. As Newton described the universe, the second body exerts an equal and opposite force on the first, which means the plaintiff and your sidewalk were in cahoots and pulled the job off together for the purpose of forcing you into a frivolous lawsuit.

What if you locked yourself in a room where no other things existed? What would happen to gravity then? Believe it or not, if you could find such a room, it would be the one place in the universe where you could go and be gravity-less. After all, there would be nothing pulling on you and you wouldn't be pulling on anything else. Of course, it's obvious that no

Going back to Newton's law, he concludes by saying that the second body exerts an equal and opposite force on the first. When we fall down the stairs, it's easy for us to feel like a victim of gravity. We might have lost our footing, but it was gravity that pulled us down all three flights, wasn't it?

Maybe not. Or not entirely. After all, Newton says that the second body exerts an equal and opposite force on the first. Could it be that as we take a tumble down the stairs we are actually pulling the ground up to meet us every bit as much as gravity is pushing us down toward the ground?

Let's not get too carried away here . . . because we also have to remember that such room exists, because to have a room, you have to have walls—and walls are just another object in the universe with which we are always having this push me–pull you interaction.

the Earth is much bigger than we are, and Newton's second law states that the acceleration of a body is proportional to 1 over mass. So since 1 divided by the Earth's mass is about equal to zero, compared with 1 divided by your mass, the acceleration of the Earth is pretty much zero compared to your acceleration. Crazy, huh?

Ever wonder why things weigh less on the moon? It's really very simple. As we said earlier, weight is really a measurement of the gravitational attraction between you and the planet you are standing on. Yet the formula for your weight on Earth says that it is proportional to the masses of the Earth and the other body (you) and inversely proportional to the square of the distance between you and the center of the Earth. Your weight on the moon is proportional to the mass of you and the mass of the moon and inversely proportional to the square of the distance between you and the center of the moon. Since the mass of the moon is 80 times less than the mass of the Earth, and since the radius of the moon is only four times less than the radius of the Earth, the gravitational attraction between you and the moon is also substantially less (about one sixth) than between you and the Earth.

THE ULTIMATE GRAVITY PROOF

Until the day he died, Newton maintained that the idea for his universal theory of gravitation came to him after an apple fell from a tree and hit him in the head.

True?

No way to tell. But this much is known: Once he put all the pieces together, he came to the conclusion that the force of gravity between two bodies should be proportional to their two masses and inversely proportional to the square of the distance between them. And armed with that chunk of scientific theory, he knew it was time to put his theory to the test, so he tried it out on two of the largest bodies he could find—the Earth and the moon. And so he would have something to compare his results to, he also tested his formula out on the Earth and his apple.

It is important to keep in mind what Newton was testing. If his theory and formula for gravity (the force of gravity is proportional to the masses of two bodies and inversely proportional to the square of the distance between them) was correct, he should have been able to plug in a few numbers (1. the mass of the Earth; 2. the mass of the moon; 3. the mass of the apple; 4. the distance between the Earth and the moon; and 5. the distance between the Earth and the apple) and get the same result both times.

Nobel Prize winner Richard Feynman did a masterful job of explaining the calculations involved, in *The Feynman Lectures on Physics*:

> We can calculate from the radius of the Moon's orbit (which is about 240,000 miles) and how long it takes to go around the Earth (approximately 29 days), how far the Moon moves in its orbit in 1 second, and can then calculate how far it falls in one second. (That is, how far the circle of the Moon's orbit falls below the straight line tangent to it at the point where the Moon was one second before.) This distance turns out to be roughly $\frac{1}{20}$ of an inch in a second. That fits very well with the inverse square law, because the Earth's radius is 4,000 miles, and if something which is 4,000 miles from the center of the Earth falls 16 feet in a second, something 240,000 miles, or 60 times as far away, should fall only $\frac{1}{3600}$ of 16 feet, which is also roughly $\frac{1}{20}$ of an inch.

Voila! Gravity!

Unfortunately for Newton, this scenario, as Feynman paints it, is not what happened.

Instead, there was a difference between the two results that was sufficiently large to cause him to abandon the entire notion.

Six years later, a group of sheepish astronomers came forward and announced to the world that they had been wrong about some previous calculations. For one thing, where they had once thought the Moon was 323 million meters from the center of the Earth, they had recently come to the conclusion that it was actually 380 million meters away. (Okay, it wasn't a huge error—only about 12 percent. But when you square the error, you come up with a 25 percent discrepancy, and that was something to worry about.)

On hearing that this figure for the distance from the Earth to the moon had been corrected, Newton calmly dug his old notes out of the desk and redid the math; this time, his old formula for gravity worked perfectly.

For more on gravity from Feynman, see his book *Six Easy Pieces: Essentials of Physics Explained by Its Most Brilliant Teacher*.

THE OTHER ULTIMATE GRAVITY PROOF

The Earth is round.

Ever wonder why?

Believe it or not, it's because of gravity. After all, if everything attracts everything else, then the center of the Earth must be constantly pulling in on everything outside of the center. And if that's the case, any corners the Earth might develop would automatically get rounded as the force of gravity pulled them in. Of course, there's more to it than just gravity . . . there's also the tendency for things to always flow toward a lower state of energy, as seen to be the case with water, which always flows downhill.

THE FINAL ULTIMATE GRAVITY PROOF: WHY THINGS ORBIT

If there were no gravity, speeding rockets would just keep going in the direction you sent them—away. But that's not what happens. When we launch a satellite, for instance, it goes up high enough to get above the Earth's atmosphere and air drag, yet in spite of its speed and even after the rocket has finished firing, it stops flying away and begins to circle the Earth. Why? Because an orbit is really a combination of two distinctly separate things: One is the tendency to go in a straight line (away from Earth, for instance), whereas the other—gravity—pulls the body back toward Earth. These two forces—in constant combination—cause the satellite to orbit around the Earth instead of crashing into it or flying farther and farther and farther away.

MARILYN, ELVIS, AND ISAAC?

Pity poor Isaac Newton. He was every bit as much a pop icon as Elvis Presley and Marilyn Monroe are today. Yet without TV or supermarket tabloid newspapers around to help spread the word, stories like these have never been told . . . until now.

NEWTON HANGS 'EM HIGH!

New Member of Mint Says
Anyone Caught Counterfeiting
Will Hang According to
Latest Laws of Science

"Let's party!" Sir Isaac Newton says, using public executions as an excuse to have a good time.

Ever wonder how Newton's universal law of gravitation might affect someone at the end of a noose?

Well, Sir Isaac used to wonder about it, too.

Newton moved to London and a good friend secured him a job at the mint. The job was meant to be an honorary one, by the way, for he is a world-famous scientist—no one expected that he was actually going to show up for work every morning. However, Newton has gone on a crusade against counterfeiters and cheats.

In addition to counterfeiting, criminals and cheats have another way of stealing from the mint—they file small amounts of gold or silver from each of their coins. Sir Isaac, however, has a solution to stop this illegal practice. The mint now flutes the edges of all the coins so there's nothing left to steal.

THREE THINGS ISAAC NEWTON AND A FIRST LADY HAVE IN COMMON

Newton and Nancy Reagan, partners in crime?

- They both shop at the same stores. Like the former First Lady, Isaac Newton went in for gowns with high, frilly collars.

- Both go in for the occult. During President Ronald Reagan's time in office it was leaked to the press that Nancy Reagan consulted with an astrologer almost daily and even used the mystic's services to help her husband make some of his presidential decisions. Newton, on the other hand, might have been one of the greatest scientists in the history of Great Britain—if not the world—yet later in his life he became fascinated with numerology, hidden meanings within the biblical book of Revelation, and other mystical pursuits.

- Both tried to hide their interest in these otherworldly pursuits. Newton, of course, was more successful at this than was Mrs. Reagan. It wasn't until his family released his diaries to the public more than 200 years after his death that this aspect of his personality was revealed.

NEW STUDY SHOWS IT'S NOT JUST CRIMINALS WHO HATE NEWTON

High school students worldwide are also said to curse him daily.

Here's why: When it came time for Newton to prove his laws of universal gravitation mathematically, he couldn't do it using the existing methods. But he knew he was right, so he invented his own system of calculation that he could then use to prove that gravity did exist.

The system of calculation he invented was readily adopted by others in the mathematical and scientific worlds and it soon came to be called calculus.

For some, calculus provides a useful and time-saving tool for dealing with an ever-changing quantity as if it were actually an infinite number of small fixed quantities. For others, it is a source of confusion and anger and frustration during their high school (or even college) years . . . and the point where their course work in mathematics ends.

(Okay, to be fair, some say that Newton wasn't the inventor of calculus, or at least he wasn't the sole inventor of it. In fact, some of his contemporaries—such as Gottfried Leibniz and Robert Hooke—had, independently, come up with their own methods for calculating changing quantities that worked just as well as the Newtonian method.)

MAYBE IT'S JUST A SIGN OF THE TIMES

Today when we want to put someone famous into office, we usually make it a TV or movie star, an astronaut, or a sports figure. Ronald Reagan was one obvious example, as well as Gopher from *The Love Boat,* who became a Republican congressman from Iowa, and Clint Eastwood, who became mayor of Carmel, California.

Back in the late 1600s, folks weren't as sophisticated as we are . . . so instead of electing their cultural icons to represent them, they elected really smart people. A case in point was Sir Isaac Newton, who was elected to Parliament at the age of 46.

To be fair, we will give Ronald Reagan, Fred Grandy (Gopher), and Clint Eastwood credit for one thing: They probably had a lot more to say as leaders then Newton did. In fact, it was only after two years of attending sessions of Parliament that Newton stood up to speak for the first time. Given his reputation as one of the greatest scientific minds of all times, the room fell silent and all eyes turned to him. *What great addition will Sir Isaac make to the current debate?* those in attendance must have silently wondered to themselves.

And they were surely disappointed when they found out that he had risen to speak only because he was cold . . . and he wanted to know if a window could be closed.

For a less amusing—actually, hard-hitting—account of Newton's life and science, see *Isaac Newton and the Scientific Revolution,* by Gale E. Christianson.

KEEPING THE PLANETS IN LINE

It's hard to believe that the same force that is there every minute of the day for you, working hard to help you keep your feet on the ground and keep your glass of wine from floating off the table, is at the exact same time also keeping all the planets true to their individual orbits. Here's how it works:

ELLIPTICALLY SPEAKING

Everyone had a different way of developing and then testing theories about the movement of the planets. Some, like Galileo and Newton, were theoretical in their methodology. They'd go outside for a while, watch the skies, then go back inside and think. Johannes Kepler, on the other hand, advanced his own observations about the movement of the planets on the basis of 20 years of observation of the skies—but the data he based his theory on wasn't his. It was information he had inherited from his boss, Tycho Brahe.

Lucky Kepler. He was at the right place at the right time, for contained in Tycho's notebooks was the most detailed account of the movement of the stars and planets that has ever been compiled. However, Tycho wasn't much on sharing. In fact, for as long as he lived, he refused to share the data he collected with anyone—but then, less than a year after Kepler went to work for him, Tycho died. And Kepler was left with the notebooks in his hands.

As Kepler analyzed Tycho's data, he discovered something that no one else

May we pause here for a little gossip about cosmetic surgery?

It is impossible (at least for us) to talk about Tycho without making some reference to his nose. Or lack of one.

Apparently, while fighting in a duel, he lost the one he began life with. To compensate, he wore a nose shield made of silver or gold (depending on which element went best with what he was wearing that day, we suppose).

had seen . . . the Earth (and all the other planets) doesn't circle the sun at a constant speed but follows a pattern of faster and slower movement. "What could cause this to happen?" Kepler asked himself. The one answer he could find mathematically that made perfect sense was that the planets were not orbiting around the sun in circles, as everyone thought, but that they were orbiting around the sun in the shape of an ellipse.

> While we are at the physicists' gossip mill . . .
>
> It is also worth noting that Kepler's life, like that of his boss, was filled with a few strange events. For instance, he once had to come to the legal defense of his mother—who was on trial for being a witch.

What Kepler knew was that ellipses have several special properties that circles do not. Whereas a circle has a single point at its center, an ellipse has two fixed points, called foci, that function similarly. You can draw an ellipse using two nails or thumbtacks, a length of string, and a pencil. Put the nails or thumbtacks into a board a few inches apart. (If you put the two nails into the exact same spot, the ellipse you draw will be a circle. In fact, the circle is just one form of an ellipse.) Tie your string in a knot so that it forms a loop. Put your string over your two nails and pull one end of it with the tip of your pencil. With your string pulled, run a loop around your two nails with your pencil.

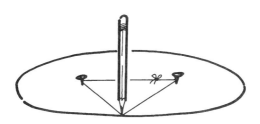

DISTANCE ÷ TIME = SPEED

Kepler's second law is this: The line from the sun to any planet sweeps out equal areas of space in equal time intervals.

Kepler was born eight years after Galileo. Like Galileo, he made some fundamental discoveries about the workings of the universe, but he, too, had no idea how to explain why each of the planets would trace the pattern of an ellipse instead of a circle. Once again, it was Isaac Newton who figured that one out and again got to claim all of the glory.

At first glance, you might think, *So what?*

Look again.

The ellipse, you know, represents one complete orbit, or one year, in the life of a planet. Therefore, the line that forms the ellipse could be thought of as representing time. It takes one year to make one trip around.

Of course, the ellipse also represents space or distance. We've drawn it to represent an orbit around the sun, after all.

Suddenly, bells go off. Time. Distance. There's a relationship here.

Yes—it's called speed, or distance traveled divided by elapsed time.

Ahh. Something's going on here. If the line from the sun to any planet sweeps out equal areas of space in equal time intervals, yet the distance covered around the ellipse in that equal time interval varies, then the speed at which a planet orbits around the sun must also vary. Why?

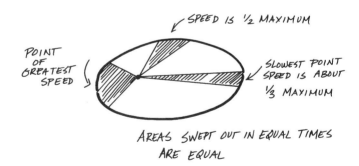

AREAS SWEPT OUT IN EQUAL TIMES ARE EQUAL

It took Kepler 10 years to figure it out.

A MATTER OF GRAVITY

For 10 long years, Kepler tried to find a relationship between the size of each planet's orbit and the amount of time it took each planet to make a complete orbit around the sun.

Here's what he came up with: The squares of the times of revolutions (or years) of the planets are proportional (or at least close to proportional) to the cubes of their average distances from the sun.

You don't have to do the math to see that what Kepler was describing was a relationship between the average speed at which planets move and their average distances from the sun. Sound familiar? Sure enough, just like everything else, the farther the planets are from the sun, the slower they move. Once again, it's the result of gravity (even though Kepler didn't know that at the time).

GRAVITY'S NOT THE ONLY FORCE

Now that we understand what's happening out there as a planet orbits, let's factor in more of what we know about gravity.

We already know that the closer two bodies come to one another, the stronger the attraction between them. Which means that if you're a planet, you're always caught up in a tug-o'-war. After all, regardless of how close or far from the sun you are in your orbit, the sun is exerting a pull on you. Meanwhile, you are also governed by the law of inertia—since you are moving, you are required to move in a straight line forever . . . unless some force causes you to veer from the path.

Obviously, planets don't move in a straight line forever, so some force must cause them to change from moving in a straight line to moving in the curved direction of an ellipse. That force is the gravitational pull of the sun. On the other hand, the planets don't stop in their tracks and plummet inward to be consumed by the sun; they still try to move in a straight line, so there must be something that resists the force of gravity. That resistance is inertia.

When it comes to the movement of the planets, inertia and gravity are locked in a knock-down-drag-out fight. Neither can get the upper hand; the resulting phenomenon is called centripetal acceleration, a type of acceleration that causes a body to move in an elliptical or circular path around a center point.

Want a real-life example?

Tie a weight to the end of a rope. Hold on to the other end of the rope in your hand and swing the weight around and around and around over your head. Easy, huh? As long as you keep hold of the string, the weight will fly in circles over your head. You, pulling on the string, are analogous to the gravitational force. But why doesn't the weight just obey you—when you pull on the string—and hit you in the face?

The answer is because of inertia.

The weight, like each of the planets, is caught in the middle of a battle between two worthy opponents—gravity and inertia.

Of course, it's not only the sun that's pulling on the planets . . . they are pulling on each other, as well. In fact, it is because of this pull that a new planet was discovered in 1846.

All planets demonstrate a wobbling effect (or perturbation, as physicists call it) from time to time when they get close to another planet, which is the effect of their pulling on each other, something that was almost impossible to explain until Newton proposed his theory of gravitation.

In the early nineteenth century, the astronomers' version of the solar system ended with the discovery of Uranus, which we think of as the seventh planet. Yet two astronomers made the observation that Uranus was acting strangely in the face of the rules that seemed to govern the other planets. Were Newton's rules wrong? Or could there be something out there in the sky with a mass large enough to pull on Uranus when the two bodies drew close?

The answer, it turns out, was that Newton's laws not only performed perfectly but they also gave the two astronomers the mathematical data they needed to locate this new planet, which we now call Neptune. And it didn't stop there. Pluto was located the same way in 1930 and, more recently (in 1998), another smaller body was discovered beyond Pluto.

Is it a tenth planet in our solar system? Only time will tell, although astrophysicists today believe the mystery body is too small to be considered a planet.

The best-selling science book ever written is Carl Sagan's *Cosmos*, and for good reason: Using his brilliant storytelling ability, Sagan unraveled the history of physics and the scientific method in a way seductive and compelling for readers at any level of scientific comprehension.

DO-IT-YOURSELF GRAVITY: YOUR HANDY GUIDE TO PLANETARY GRAVITY STATS

Planet	Closest Distance to Earth	Average Distance from Sun	Average Orbital Speed	Time Span of One Complete Orbit Around Sun	Relative Mass (When Earth = 1)
1. Mercury	0.54 au*	0.39 au	29.76 mi/sec†	88.00 days	0.060
2. Venus	0.27 au	0.72 au	21.75 mi/sec	224.70 days	0.820
3. Earth	Not applicable	1.00 au	18.52 mi/sec	365.30 days	1.000
4. Mars	0.38 au	1.52 au	14.98 mi/sec	687 days	0.110
5. Jupiter	3.95 au	5.20 au	8.14 mi/sec	11.86 years	317.900
6. Saturn	8.00 au	9.54 au	5.98 mi/sec	29.46 years	95.100
7. Uranus	17.28 au	19.18 au	4.23 mi/sec	84.01 years	14.600
8. Neptune	28.80 au	30.06 au	3.36 mi/sec	164.80 years	17.200
9. Pluto	28.72 au	39.36 au	2.92 mi/sec	247.70 years	0.002–0.003

And just for fun we're throwing in the stats on the sun and the moon, too:

Sun	0.98 au	Not applicable	Not applicable	Not applicable	333,000.000
Moon	0.0024 au	1.00 au	0.621 mi/sec	Not applicable	0.012

* An "au" is an astronomical unit, defined as the mean distance between the earth and the sun—92,958,348 miles, 149,597,870 kilometers, or 499 light-seconds.

† mi/sec = miles per second

GREAT STRIDES IN PHYSICS

"Go home—all the mysteries of the universe have already been solved." By the end of the nineteenth century, physicists of the classical school were pretty darn proud of all they had accomplished . . . so much so that they began to turn students away and brag that they had successfully discovered everything there was to know about the physical universe.

To their credit, physicists made great strides once Newton unveiled his three laws. In addition to their understanding of the old standbys—motion, gravity, and the movement of the planets—they also made dramatic discoveries about the mysteries of mechanics, electricity and magnetism, heat, sound, and light.

CLASSICAL MECHANICS

Once Newton unveiled his three laws and they became universally accepted, it seemed to physicists that there wasn't anything mechanical the laws couldn't explain. The world, it seemed, had become as predictable and as easy to understand as a clock.

The genius of Newton's laws was that they made it possible to predict the behavior of anything mechanical that the eye could see . . . reducing the complex world to a simple machine much like a car, a bike, or an eggbeater.

The Way Things Work Around Here

Mechanics is the study of objects in motion . . . and when we talk about objects in motion, one of the things we are talking about are machines. But we're not talking about the traditional definition of the word *machine.* Instead, when we say "machine," we mean any device that allows us to perform some "work" by combining elements of force and motion.

Take your car, for example. Not only is it a machine that allows you to get from place to place fairly effortlessly, but if you look under the hood you will see that it is actually a large collection of smaller machines that, when combined, make it possible for the bigger machine (your car) to do its job.

Until the advent of the computer, all machines were mechanical. (We don't mean to say that computers aren't machines; it's just that they are electronic machines instead of mechanical ones—the difference is that electronic machines substitute electricity for motion.) What makes every mechanical machine work is a collection of pistons and valves and gears and rods and belts that move and work together to complete some designated task, whether it be getting you from point A to point B (a bike or a car) or whipping up some eggs (an eggbeater).

Regardless of how it works and what it does, if a machine is mechanical its roots can be traced back to the six most basic forms of machinery: the lever, the inclined plane, the wheel and axle, the pulley, the wedge, and the screw. Furthermore, all mechanical machines have these two properties in common:

- **Machines don't do their jobs until you do yours.** Regardless of what task a machine performs, yours is always the same: You're the one who's got to provide the machine with energy, whether it be in the form of your own effort, the effort of an animal, or some form of fuel that you supply. Wish you could find a machine that would feed itself? Sorry, but that would be called a perpetual motion machine and the laws of physics say such a thing can't exist. Which brings up an interesting point: A machine is worthwhile only if it costs you less in energy to use to complete a task than it would for you to do the same task by some other means. Sounds complicated, but it's not. If you want to travel from Los Angeles to New York, for instance, you have several machines at your disposal to help you do the job. You can fly. You can take the train. You can take the bus. You can drive your own car. You can ride a bike. You can walk. (Yes, you, too, are a machine.) For most people, flying is the clear choice because regardless of the airfare (which is another way of supplying the energy necessary to make a machine go), flying still takes a lot less of your energy (and time) than do the alternatives. This example is good because it points out that the way in which we pay for a machine to do work for us can vary drastically. In some cases, our input takes the form of our own sweat and blood, but in other cases it may take the form of a stockpiled resource, like money or wood or a fossil fuel.

> **ALTERNATIVE ENERGY SOURCE?**
>
> After a day of long, hard toil in the summer sun as a boy, one of your authors asked his father why he always had to do so much yard-work. His father's response: "Why do you think I had sons?"

■ **All machines are governed by an efficiency ratio**, but because of energy lost to friction, heat (like the exhaust from a clothes dryer or your car), and other factors, this ratio is always less than 100 percent. Even when the machine in question is you, the efficiency ratio is always less than perfect—which explains the sweat dripping off your brow after you help a friend carry a refrigerator up three flights of stairs.

AN ANALOGY TO DROOL OVER

You could say that mechanics is to physics what hamburgers are to McDonald's.

Sure, Ronald McDonald's got his shakes and his fries and his apple pies, but it's those darn burgers that make it all possible—without them, the McDonald's universe wouldn't exist.

Same thing with physics; you've got your relativity and your quantum theory and your quarks and your black holes, but it's a few basic rules about the way things work around here (a.k.a. classical mechanics) that define the everyday universe as most of us know it.

Here's another thing that physics and your local McDonald's have in common: uniformity.

Have you ever noticed that when you go into a McDonald's, the menu is the same as every other McDonald's? The counter where you place your order is the same. The wrapper around you burger is the same, and—amazingly—even the burger itself is the same, no matter what.

Here's why: The owners of McDonald's have discovered something that physicists have known for a long, long time. A universe (whether it's a universe the size of our own or simply the McDonald's hamburger universe) is governed by a few ironclad rules. Furthermore, once you understand what those rules are, how they work, and their repercussions, you can begin to use them to work for you, instead of being at their mercy.

Let's stick with the McDonald's analogy: If we, the owners of McDonald's, make an ironclad rule that every single cheeseburger we make, regardless of where it is made or who makes it, should be perfectly round, contain eight ounces of meat, be cooked for three minutes in a 400-degree broiler, and served in 240 seconds or less after leaving the grill, we will know that no matter what McDonald's a customer walks into, he or she will have a McDonald's cheeseburger that exactly looks and tastes and smells like the McDonald's hamburger the customer has come to know and trust.

How far does McDonald's go to make sure you have a uniform and consistent McDonald's experience? Well, all the meat comes from centrally located processing plants, as do the buns, the lettuce, the catsup, the pickles, and so on. The cooking itself is a uniform process because the cooking equipment in each McDonald's restaurant is exactly the same. Finally, whether you go to a McDonald's in Boston or Bakersfield or Biloxi, you will even be greeted the same way: "Welcome to McDonald's. May I take your order, please?"

But what does this have to do with physics?

Much like McDonald's wants you to have a uniform experience by creating a system of rules of laws that govern behavior, Mother Nature has done the same thing. She wants you to have a uniform experience of the universe no matter where you are, so she made the same rules apply, no matter where in the universe you may go.

In the case of McDonald's, it's easy to see that what the company has done was turn each of their restaurants into a duplicate, well-oiled machine, replicating itself over and over again as it turns out billions and billions of Big Macs, each just as much like the very first one as the others. In many ways, our universe works the same way.

There is one big difference between the system or rules that govern a McDonald's and the system or rules that govern our universe: In the case of McDonald's, the company was free to create a burger universe that fit its needs and then tinker with the rules along the way to make things run even better. The opposite, of course, is true when you're dealing with Mother Nature. She's already made the rules, and the best we can do is hope to get better and better at understanding them.

"Whattdaya Think I Am . . . a Machine?"

Yup, we sure do. In fact, some of the best machines known to humankind have had you as one of the parts:

You as a Lever

You know what Archimedes said: "Give me where to stand, and I will move the earth."

Whether you are cutting paper with a pair of scissors, tearing down a wall with a crowbar, or playing on a teeter-totter with one of your kids, every time you use a lever the principle is always the same: You have a rigid beam positioned over a fulcrum so that when you apply force on one end of the beam, it will move the load at the other end—and the closer you put the load to the fulcrum, the easier your job will be. In the case of a beer curl, shown in the illustration at the right, your wrist provides the energy, your

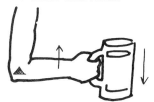

elbow is the fulcrum (1), and your arm (3) is the lever. The beer (2), of course, is the load.

You as a Wheel and Axle

Technically speaking, the wheel and axle are really just another example of a lever, with the axle acting as the fulcrum.

Some might think of the wheelbarrow as the perfect example of a wheel and axle, for two reasons: first, because the wheel allows you to roll your cargo instead of dragging it, thus reducing friction, and second, because the wheel acts as a fulcrum and makes it easier for you to lift your load from the back of the barrow. But a better example—with you as the star—is your wrist (the lever) and arm (the axle) when you open a jar of nuts.

You as a Pulley

Much as the wheel and axle is a specialized form of lever, the pulley is a specialized form of wheel and axle in that it is secured at a fixed point.

The simplest pulley consists of a fixed wheel and a rope that loops around it a single time. A more complex pulley (called a block and tackle) has multiple wheels around which the rope makes multiple loops.

Here's why a pulley (or block and tackle) is so useful: For each loop a rope makes around a wheel, the amount of distance required to move the rope is doubled but the amount of force required to move it is halved. It still requires the same amount of work to get the job done, however. With each pull of your arms on the rope from above your head down to your waist, only a small amount of lift is produced on the other end. You do more pulling overall, but each time you pull down, less force is required of you.

The bottom line: If your block and tackle has two wheels (therefore, two loops), you can lift 100 pounds with the same amount of effort it would have taken you to lift half that weight with a single pulley alone.

Although most of us don't have many opportunities to step in and allow ourselves to act as pulleys or blocks and tackle, the benefits of this technology will become apparent to you should you ever be the last person on the rope in a tug-of-war and you make the mistake of wrapping the rope around a tree before you wrap it around your waist.

You as an Inclined Plane

Every once in a while, something in physics is so mathematically simple that it would be a shame not to explain it. A case in point is the inclined plane.

To best understand why it works, imagine you have a pickup truck and need to put a 100-pound box in the back. You have two choices: (1) you can lift the 100-pound box high enough to get it onto the tailgate and pray you don't ruin your back or (2) you can measure the height of the tailgate and then find a piece of plywood that is twice as long. Put one end of the plywood on the ground and the other end on the tailgate and slide your heavy box up your inclined plane.

Not only is it much easier to get the box into the truck this way but it is also a cinch to understand, mathematically, because a plane that is twice as long as the height of the tailgate will require exactly the same work. Or, stated another way, an inclined plane will always reduce the force in direct proportion to the increased distance traveled.

Meanwhile, if you ever find yourself in a large city like New York and you are in need of work, you might consider applying for a job as a piano humper. These are people whose job it is to act like an inclined plane and hoist heavy furniture onto their backs and carry it up the stairs to apartments on high floors in buildings without big enough elevators.

You as a Wedge

A wedge is a special form of inclined plane that is usually made of strong material (usually steel or wood) and can be driven into that which you want to hold or split.

The head of a hatchet is a wedge, a chisel is a wedge, and a doorstop is a wedge—and so are you when you jam your arm between the doors of an elevator to get them to stop.

The wedge is effective because when you apply a forward or downward force to it, that force is translated into a strong sideways movement against whatever it is the wedge is wedged in between.

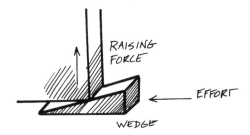

RAISING FORCE

EFFORT

WEDGE

You as a Screw

If you took a good long look at a wood screw, you would see that it is actually a very long inclined plane wrapped around a central shaft. Try as we might, we can't think of a single example where the human body functions in the same way.

If you want to learn more about basic machines without ever breaking a sweat, see Michael S. Dahl's excellent books on the subject: *Inclined Planes, Simple Machines, Levers, Pulleys,* and *Wheels and Axles*. If you want to know how almost every machine you can think of works, see David Macaulay's *The Way Things Work*.

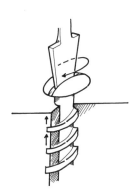

ELECTRICITY AND MAGNETISM

Think of electricity as gravity with a split personality.

Gravity attracts, whereas electricity both repels and attracts. And the two forces work similarly: Gravity attracts with a force that is proportional to the mass of two bodies and inversely proportional to the distance between them, and electricity attracts or repels with a force that is proportional to the charges of two bodies and inversely proportional to the distance between them.

Isaac Newton brought about a revolution in the field of mechanics. A revolution in the area of electricity was not far behind; one of the men leading the charge was none other than Benjamin Franklin.

Humankind's knowledge of electricity dates back several thousand years to ancient Greece when a guy named Thales (625–545 B.C.E.) observed that if

> ### · · · BUT DON'T CONFUSE ELECTRICITY WITH ENERGY
>
> You'll recall that early on, we defined *energy* as any force that can produce a change in matter. Electricity, then, is just one form of energy, along with others, such as light, sound, and heat.

he rubbed a chunk of amber—which the Greeks called ἤλεκτρου, or *elektron*—with a cloth, the amber would then attract tiny pieces of straw. Many years later, it was observed that if one rubbed a glass rod with a piece of cloth or rubbed a piece of sealing wax with some fur, the rod and the wax would suddenly have the ability to pick up small pieces of straw and even paper.

It was Franklin who concluded that all bodies carry within them a collection of electrical charges, which he called negative and positive. Under normal circumstances, Franklin said, all the charges in a body will be equal—in other words, within a body there will be just as many positive charges as there are negative, so these charges will cancel each other out and the body will be neutral.

However, under certain conditions, that balance of charges could change and a body could have more of one or the other. Such is the case when we rub glass with a cloth or wax with a piece of fur. The glass becomes positively charged when rubbed with a cloth, and the wax becomes negatively charged when rubbed with fur.

Suddenly, the reason for two rubbed pieces of glass to repel one another becomes clear—both pieces have electrical charges that are out of whack, out of balance. Both pieces of glass rod are negatively charged. Meanwhile, we can now explain why the rubbed glass attracts the rubbed wax. One is negative and one is positive. And with electricity, likes repel and opposites attract.

Inasmuch as discoveries in physics often come down to applying the same laws and rules and patterns over and over and over again to different sets of phenomena, it is interesting to note that there are sometimes patterns in the human drama behind these discoveries that are repeated as well. For instance, Ben Franklin, like Galileo and Kepler before him, was able to predict what would happen given a certain set of circumstances, but he was never able to explain why these things happened.

FOUR THINGS BEN FRANKLIN NEVER KNEW ABOUT ELECTRICITY			
Where Does the Charge in a Body Come From?	If Matter Is Normally Neutral, How Come It Sometimes Gets a Charge?	How Come Some Materials Seem to Pass On (or Accept) Electrons Better Than Other Materials?	Since Likes Repel and Unlikes Attract, Does That Mean That Electricity Is the Same Thing as Magnetism?
All matter is made up of atoms and all atoms are made up of protons, neutrons, and electrons. All electrons carry a negative charge and all protons carry a positive charge. Under normal circumstances, an atom will have an equal number of protons and electrons, so that the atom itself is neutral. However, when the balance of those electrons and protons gets out of whack, the body becomes electrically charged.	Some atoms have the capacity to hold or attract more electrons on a temporary basis than they have protons to neutralize them. When this happens, as in the case of the cloth we used to rub the glass rod, the body takes on a negative charge. At the same time, those negatively charged atoms had to come from somewhere, and in this case they came from the glass rod. Because the glass rod gave up some of its electrons, it now has a	Some objects—like our glass rod or a hard piece of rubber or our piece of wax—do not give up their electrons easily. These are called insulators. Other objects, like most metals, are called conductors because they allow electrons to pass through them easily. As for why one material, like metal, is a good conductor whereas another material, like cloth, isn't—the answer lies in the composition of the atoms within each. The electrons in the conductor—metal,	Uh, oh. Now you did it. You went and asked an embarrassing question. You see, physicists have known about the existence of both electricity and magnetism for many years, but it wasn't until the mid-nineteenth century that anyone figured out that they are actually the same force with two different external manifestations, which we now call electrodynamics.

Where Does the Charge in a Body Come From?	If Matter Is Normally Neutral, How Come It Sometimes Gets a Charge?	How Come Some Materials Seem to Pass On (or Accept) Electrons Better Than Other Materials?	Since Likes Repel and Unlikes Attract, Does That Mean That Electricity Is the Same Thing as Magnetism?
	positive charge because it has more protons than electrons.	for example—are very loosely bound and are free to wander about the boundaries of the entire structure. (This structure might be a cookie sheet you have just put in the oven or it might be a strand of copper wire that runs from a power plant all the way to your house.) However, within the insulator—cloth, for example—each electron is assigned to a particular atom instead of being free to travel at will. This explains why electricians use wire (because metal is a good conductor) to move electricity from one spot in your house to another, whereas they wrap the wire in rubber, which is a good insulator, to keep from getting shocked.	

"Hey! You got electricity in my magnetism!"

"Hey! You got magnetism in my electricity!"

Introducing electrodynamics, the Reese's peanut butter cup of physics.

THE ONE BOOK I WOULD READ . . .

When I was a child, perhaps the books I read most eagerly and enjoyed were books by Earnest Thompson Seton like *Two Little Savages*—it's about two boys living outdoors and learning how to take care of themselves. They were role models for me as a boy, because I liked to do just those kinds of things.

I built a tent and stayed out in the woods awhile and my family thought it was just fine for me to explore nature and learn how to take care of myself—I would stay overnight and have my tent as just kind of a little visiting and work place.

I think, as an adult, the book that I could say probably influenced me more than any other was a book on electricity and magnetism by Smythe. [He] was a professor with whom I was doing graduate work. I went over it very carefully with him while he was writing it; I helped him and worked all the problems and so on, so that it was a field I came to know very, very thoroughly. It's an important field and this book has been an important resource for me.

—Charles Townes, winner of the Nobel Prize in physics in 1964

For an excellent—and more up-to-date book—on electrodynamics, see *Introduction to Electrodynamics,* by David J. Griffiths.

- Just as with electricity, like ends of a magnet will repel each other.
- When you pass an electric current through a wire, it will cause a magnetic field to be generated around the outside of the wire. (Interestingly, the magnetic field will always flow in a right angle to the direction of the current inside the wire.)

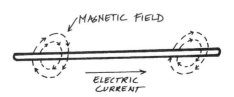

▪ Oh, but what if you do the opposite—if you pass a magnet through a wire loop? If you wrap some wire around a toilet tissue roll tube and then push a magnet through the tube, you will create some electric current.

HEAT

Think of it as the Fresno of physics—Fresno, California. It's one of those places where you'd never go if it weren't on the way to someplace else.

In this case, someplace else is Yosemite National Park, a place everyone wants to go. In fact, close to a million people a year from all over the world fly into Anaheim to see Disneyland and Hollywood and Knott's Berry Farm and then climb into a rental car and drive six hours north to Yosemite. And it just so happens that while they're on their way to Yosemite, Fresno is the last place they can stop to go to the bathroom before they begin their ascent into the mountains toward the park. So even though they're on way to Yosemite and don't have any interest in visiting Fresno . . . since they've gotta go, Fresno is where they stop.

In the realm of classical physics, heat plays a similar role—it's not really where we want to go, but it's sort of on the way to where we're headed.

The Three Faces of Heat

1. When you talk about heat, things quickly become confusing—because to a physicist, the word *heat* has many meanings, but none of them are the meanings that most people normally associate with it. When most of us hear the word *heat* we think of the weather outside in the summer or the dollars and cents we spend on astronomical gas bills in the winter.

   Physicists, on the other hand, have a different take on the word. Heat, to them, believe it or not, involves change. In other words, if you take some matter and change it, you not only get something different—you also get some heat.

2. When you talk about heat, what you're really talking about is the speed at which molecules move: In an object with a

high temperature, molecules move quickly, and in an object with a low temperature, molecules move slowly.

3. When you talk about heat, what you're really talking about are the laws of thermodynamics, which is another name for the set of laws that govern the way in which objects of different temperatures will interact when they come in contact with one another.

The Three Other Faces of Heat

4. First, we've got heat[1], the textbook definition: Heat is energy that passes from one body to another when there is a temperature difference between the two. According to the laws of thermodynamics, energy will always flow from the body with the highest temperature to the body with the lowest temperature until the temperatures of both bodies are equal.

5. Then we've got the effects of heat[2]—"Ow! That's hot!"—like when you burn your finger with a match.

6. Finally, we've got the source of heat[3], which you apply to matter to effect a change, which in this case would be the match you burned yourself with.

To illustrate the difference among these three types of heat, let's say we light up the family blowtorch and apply it to that old car that's parked out in front of the house.

First, We Light the Torch

To light the torch, we've got to have a little heat[2] (courtesy of a match, let's say) to get our source of heat[3] (the torch) going. Normally, we think of this process as working the other way around: If you want to light a cigarette, what you do is strike a match, right? The problem with this picture is that it overlooks the fact that before your match (heat[3]) can become a source of heat[3], a source of heat[1] must be applied to it, just as before our torch can become a source of heat[2], a source of heat[1] must be applied to it.

Okay, so now we see that the torch and the match have a lot in common. Both must go through the same process before they can do the job for which they were created. But if we need the match to light the torch, where are we going to get a source of heat to light the match?

Actually, friction is just one of six places we can go to get heat. The others are

- The sun
- The center of the Earth (that is, the heat of a volcano or geysers)
- Chemical reactions
- The nucleus of the atom (a.k.a. nuclear energy)
- Electricity

How about if we provide the heat[1] that the match needs to start by adding some friction? After all, matches don't start spontaneously—we strike them to get them going. Although most of us have never thought about it before, the act of striking a match is similar to rubbing our hands together briskly on a cold day. Thanks to friction, our hands get warmer. Thanks to friction, the head of the match gets hot. And the chemical compound on the head of the match doesn't need much heat to ignite.

Of course, there's still a problem here. Matches don't just spontaneously strike themselves. It takes some heat[2]—some fuel, some energy—to cause a match to be struck. In this case, the source of heat[3] is you. And you don't just spontaneously burst into match-striking activity, either. You require heat[3] to do so. So you eat to provide the fuel or energy or heat that you need to strike a match.

And so on.

Now That We've Gotten the Torch Lit . . .

Wow! That flame is hot! Now, just for fun, how about if we turn the torch onto the front left fender of the car? Isn't it great how the heat of the torch causes the paint to blister and peel away (heat[2]) and then the entire fender to get red-hot (heat[3])?

Sound familiar? Well, that's the interesting thing: As we torch our car, the car goes through the exact same process we just described above. Given enough heat (heat[2]), darn near anything will ignite (heat[3]).

But that's really just half the story. Consider now that all we have learned about heat—from your eating enough to supply the fuel to strike the match to the lighting of the torch to the application of the torch to the car—as looking at heat through a telescope. In other words, now you've got the big picture. But what would we see if we were to look at heat through a microscope?

Now, Let's Get the Little Picture

- All matter—whether it be a match, a torch, the fender of a car, or you—is made up of molecules. And when you add heat (heat²), those molecules start moving around faster and faster.

- Before we move on, let's consider the following for a moment: Liquids, solids, and gases are all made up of molecules. So why are they so different? (Or why would you rather get hit with a water balloon than with a block of ice?)

- The difference between them is due to the competition between the force that holds them together (the attraction between molecules) and the force that pushes them apart (kinetic energy). (That's kinetic energy versus potential energy. Kinetic energy is the energy of motion. When engineers dam a river and then release water to turn turbines to make electricity, they are harnessing kinetic energy. On the other hand, the water that doesn't get released from the dam can be said to have potential energy by virtue of its position and its potential to spill from the dam. Another example of an object with potential energy is a loose brick at the top of a wall. The fact that it is an accident waiting to happen is a testament to its potential energy.)

- But let's get back to our torch: One form of kinetic energy is heat. So when we add heat to a solid, we increase the kinetic energy and the molecules move farther apart and move faster and we get a liquid. We can add heat to a liquid and turn it into a gas the same way.

- When molecules increase their rate of movement, there are consequences. For instance, when the molecules in many solids start moving faster, you get a liquid. When the molecules in many liquids start moving faster, you get a gas. When the molecules in many gases start moving faster, you often get an explosion.

▪ As we said earlier, heat to a physicist implies change. And it is on the molecular level where this change occurs—even though we can't see it with the naked eye until so many molecular changes have occurred that the paint starts blistering and peeling off your car.

▪ When you add heat to matter, things start happening one right after another:

1. **There is a change in temperature.** Somehow, this seems a bit anticlimactic when you're talking about blowtorching a car, but this is only the first step. After all, sometimes a change in temperature is all you really want . . . like when you're heating up some soup.

2. **There is a change in size.** When you've got all these molecules getting hotter and hotter, and thus moving around more and more and more, they're bound to need more room, right? This is particularly true with gases, which bombard whatever container you're keeping them in harder and harder and faster and faster, thus building up the pressure inside. Liquids and solids get bigger when they're heated, too, but solids expand the most.

WANT TO SEE THE PRINCIPLE OF EXPANSION IN ACTION?

Go look at a thermometer or at the sidewalk. As the liquid in the bulb of the thermometer gets hotter, it expands and shoots up higher and higher on the scale. And . . . have you ever wondered why sidewalks have cracks? It's no accident—engineers put them there so the asphalt or concrete has space to grow into on hot days.

---

3. **There is a change in state.** Now you've done it . . . you've applied so much heat to the fender that you've changed it from a hard chunk of metal into a hot stream of liquid steel. And if you could get that liquid steel hot enough, it would eventually turn into a gas. For a more familiar example of a change of state, consider what happens when you heat a chunk of ice: The ice melts and becomes water. And if we continue to add heat, that water eventually vaporizes and becomes steam, a gas.

One of the best books around on thermodynamics is *Fundamentals of Classical Thermodynamics,* by Gordon J. Van Wylen and Richard E. Sonntag.

"Is That a Solid in Your Pocket or Are You Just Glad to See Me?"

If you're going to talk about change, you've got to know what you've already got—otherwise, how will you know when it's different?

In the worlds of physics and chemistry, it is not always clear whether an object is a solid or a gas or a liquid. That's what caused Friedrich Wilhelm Ostwald, known to many as the father of physical chemistry, to conclude that you can figure out what just about anything is by comparing its physical properties to the physical properties of something else that is known to you.

Ostwald was also a big believer in the honor system, for he never said just how many properties you had to match up before you could draw your conclusions.

GAS, LIQUID, OR SOLID?				
	Shape and Volume	Pressurization	Density, Compression, and Expandability	Willingness to Combine with Like Matter
What's a gas?	Gases have no shape and no definite volume. When you stick a gas into an empty container, it will fill whatever space is available.	When captured in a container, a gas will exert a uniform pressure against all container walls.	Gases are much more compressible and expandable than either liquids or solids. Usually, gases have a much lower density than do liquids or solids.	Under most circumstances, one gas will mix freely and easily with another after accounting for differences in density and mass between the two.
What's a liquid?	Liquids have a definite	The pressure a liquid exerts	Under normal circumstances,	Different liquids will

	Shape and Volume	Pressurization	Density, Compression, and Expandability	Willingness to Combine with Like Matter
What's a liquid? (cont.)	volume but no definite shape. When you pour a liquid into an empty container, it goes to the bottom and assumes the shape of the container.	against the walls of the container will remain constant over the surface of the container and will also depend on the density of the liquid, the size and shape of the container, and how much liquid is in it.	liquids will have a much greater density than will gases. Liquids are much more compressible than are solids.	mix with varying degrees of success. Some, like scotch and soda, will mix completely. Others, like oil and water, will mix with only a limited degree of success.
What's a solid?	Solids are things that have definite shape and definite volume. When you stick a solid into an empty container, it goes to the bottom and keeps its own shape.	Under normal circumstances, the only pressure that a solid will exert on a container will be due to its own weight and mass.	Solids normally have a greater density than either gases or liquids. Solids are much less compressible and expandable than are gases or liquids.	As a general rule, solids do not mix.

Boiling water with ice! Bottled hiccups! A marshmallow in a syringe! The best way to learn about the different states of matter is to act like a kid and take a hands-on approach. To that end, we recommend *Investigating Solids, Liquids, and Gases with Toys: States of Matter and Changes of State,* which was funded by the National Science Foundation and written by Jerry Sarquis, Lynn Hogue, Mickey Sarquis, and Linda Woodward.

Back to Work, Slave!

If it hasn't become obvious yet, then let us state it clearly and plainly here: Heat is our slave. We capture it, we put it to work; tirelessly, it does our bidding. Then, we do everything we can to keep it from getting away.

You do it. We do it. Even liberal Democrats do it. Sounds horrific, but our only defense is that given the laws of thermodynamics, it's all perfectly legal—and besides, everyone else is doing it, too.

The four laws of thermodynamics:

1. **You can't win.** Law 1 is the law of conservation, which tells us that we can never create or destroy heat (or energy); we can only harness it for a little while, and when we do, some of it is always lost. (If you add the amount of energy used to do the work plus the amount of energy lost, it will always equal the total amount of energy applied.)

   Take our blow torch, for example. When we light it, most of the fuel burns in the flame, but some is lost in heat to the atmosphere. The amount of work we get from that fuel is a function of how hot the flame is and how long it burns. In a perfect world, we would light our torch and all of the heat in the fuel would be transferred to the fender.

   Instead, we lose some energy by inefficient combustion, some to light, some to sound, and some to the inadvertent heating of the torch nozzle and the air surrounding the torch. Although it is true that over time technology will increase the efficiency by which a torch burns fuel, there will always be a limit to how much work we can get out of a unit of fuel because, as the law of conservation says, we can never get more out of it than we put in.

2. **You might break even.** On a cold day, law 2 seems obvious: When two bodies come in contact with one another, heat from the warmer body will naturally flow to the colder body until an equilibrium is reached. (Anyone who has ever disobeyed his or her mother and gone out on a cold day without a coat has experienced this firsthand.)

   If you want to reverse the process—chill a warm body—then you can't count on nature to do it; instead, you'll have to add some fuel and do some work. For instance, if we have a six-pack of warm beer but know we'll want a few cold ones when we're done hacking away at that old car, we'll have to plan ahead and put the six-pack into the refrigerator. The refrigerator is able to turn the warm beer into cold beer because it uses energy. Notice, too, that although the refrigerator burns fuel to turn the warm beer cold, it also obeys the first law of thermodynamics. Most of the fuel is used to cool the beer, but the energy has to be transferred somewhere (like the coils under or behind your fridge) to be thrown off as heat—but the total of that thrown-off heat and the work done to cool the beer, by necessity, will always equal the amount of energy used to power the refrigerator. (Okay, not exactly. The truth is, you always lose a little bit of energy to entropy. Entropy? Yes, entropy. No matter where you go or what you do in life, by virtue of going or doing, you are creating some disorder, and disorder always eats up a little bit of energy. This is why a perpetual-motion machine will always be a dream and never a reality.)

3. **You can't get out of the game.** Not only do the laws of conservation tell us that we cannot create or destroy energy, but they also tell us that we cannot create or destroy matter. Law 3 bridges those two principles by stating that it is impossible to ever reach a temperature of absolute zero, the point at which molecules will cease to move. If law 3 weren't true and absolute zero *could* be attained, any matter that reached a temperature of absolute zero would be destroyed—which this law tells us we cannot do.

0. **You could have a tie.** Although this final law was recognized long after the first three, it is so basic to the principle of thermodynamics that it has been named the zeroth law. The zeroth law states that if two bodies have the same temperature and are in equilibrium with a third body, then the two must also be in equilibrium with each other.

SOUND

When Mick Jagger sets out to serenade his audience of 500,000 with a rousing rendition of "Get Off of My Cloud!" and when

Luciano Pavarotti opens his mouth to seduce his audience with his version of *La Bohème,* both open their mouths to belt their respective tunes and cause their vocal cords to vibrate. This is good old-fashioned force, by the way. When those vibrations leave the artists' mouths, they travel through the air as waves. And when the waves reach our ears, they make tiny bones in our eardrums vibrate and we hear the vibration as sound. So sound, really, is just a by-product of vibrations created by force.

As you might guess, the laws of physics that govern the behavior of a vibration are the same laws already discussed that govern all other forms of matter and motion, but when we start talking about waves, we open a Pandora's box of fascinating phenomena.

CATCH A WAVE! ELEVEN THINGS YOUR LOCAL SURFER CAN TEACH YOU ABOUT PHYSICS

Given the similarities between sound waves and waves at the beach, it's no wonder the Beach Boys were such musical geniuses.

- The highest point of a wave is called the crest. The lowest point is called the trough. Gnarly, dude.

- To calculate the wavelength of a series of waves, measure the distance from the crest of one wave to the crest of the next. (Actually, you can measure from any point on one wave to the same point on the next wave to measure wavelength. However, beginning and ending with the crest or trough seems to make the job go a little easier.)

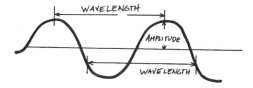

- The amplitude of a wave is half the distance from the lowest point to the highest point. It is the height of a crest or the depth of a trough.

- Frequency refers to the number of waves that will pass a given point per second. For instance, when a radio station identifies itself as "107.5 on your FM dial," what the station is really telling you is the frequency of its radio signal or that it broadcasts its signal at 107.5 megahertz, or 107,500,000 waves per second.

- If you want to know the speed of a series of waves, multiply the frequency of the series of waves by the wavelength. This will give you the wave velocity.

- You can learn a lot about waves by tying a rope to a doorknob and giving it a good short shake. For instance, you'll notice that the waves in the rope first travel in one direction toward the doorknob and then reverse direction and come back in the direction of your hand. When the medium (in this case, the rope) is perpendicular to the direction of its own velocity, it is called a transverse wave. (Keep in mind that the magnitude of velocity is the same thing as speed—where velocity differs from speed is that it also includes the direction of movement of the body. In this case, your rope runs horizontally—on average—from your hand to the door, and the wave moves and your waves move vertically, or up and down.) This is much like the string of a musical instrument; when it is plucked, the result is a transverse wave.

- In contrast to the transverse wave is the longitudinal wave, where the wave's velocity is parallel to the direction of the disturbance. The best way to illustrate this type of wave is to imagine that you have a Slinky and you are allowing it to stretch and contract. As you do this, the disturbance (or wave) runs along the length of the Slinky itself.

- The rule of superposition comes into play when two waves meet. Waves that come together in a particular place add or subtract just like numbers. For example, if the trough of one wave meets the crest of another wave, the trough will cancel the crest and the result will be a smaller wave (or no wave at all), kind of like $3 - 2 = 1$, or $2 - 2 = 0$. Or, if two crests meet, the result will be a bigger wave whose height is the sum of the two waves. When two waves meet and the result is a bigger wave, this is called constructive interference. If, on the other hand, the result is a smaller wave, it is called destructive interference.

- After a surfer paddles out past the surf and stakes his claim, he will often remain in one stationary position in the water, waiting for a really good wave—paddling just enough to stay in one spot. When this happens, his paddling motion will send out ripples, rings of waves in concentric circles with him at the center—and the distance between the crests of these waves will be equal. However, if another

WATER WAVES

A   B

surfer nearby decides that she'll get better waves if she just moves a short distance down the shore to the left and then slowly paddles in that direction, something interesting happens for anyone who is observing: As long as she paddles along at a rate that is slower than the speed of her waves, the waves in front of her will have a higher frequency than will those behind her. However, if the surfer paddles hard enough, she will begin to catch up with her own waves, so an observer at point A sees waves that are higher in frequency than the waves seen by an observer at point B. When the frequency of a series of waves is changed by the motion of the force creating the waves (a.k.a. the surfer) relative to the waves themselves, the resulting phenomena is called the Doppler effect.

- Let's say one of our surfers decides that the best waves are certain to be far down the beach. If that were the case, he would start paddling on his board as hard as he could, and if he succeeded at paddling hard enough to move himself and the board along the water at a speed that was faster than the waves he was creating, he would at some point have to overcome a wave barrier larger than any of the single waves he generated. The wave barrier would be the cumulative result of all the waves he created by paddling—but if he was able to pass the barrier, he would encounter smooth water, for he would be paddling faster than his own waves and be able to stay ahead of them. When a jet airplane crosses the wave barrier of its own sound waves, then cumulation of all of those sound waves reaches points on the ground at the same time and the result is a sonic boom.

- Finally, let's talk about the medium in which a wave exists. In this case, the medium is the ocean. Waves roll in; waves roll out. But what is the water doing all this time? You may recall that earlier we said that a vibration exists as a function of time—it is a periodic motion. The wave, on the other hand, exists as a function of space; in other words, it must move from point A to point B. But the water is neither wave nor vibration. It is a medium and it stays in the same place. This sounds far-fetched until you fill up the tub, throw in a rubber duck, make a wave with your hand, and test this out for yourself. The wave you make moves to the other side of the tub. But what happens to the duck? It bobs up and down but (basically) stays in the same place.

LIGHT

If you haven't noticed yet, Newton's discovery of gravity gave scientists the ability to coast for a few hundred years. After all, anything they discovered in this mechanical, clocklike universe was only the reapplication of the same old laws of Newton. Mechanics is pure Newton . . . forces acting on bodies. Electricity? More forces. Magnetism? Forces. Heat and sound? Even more forces.

In fact, light works the same way. Or at least that's how Newton saw it. Light begins life as a particle. Particles are matter. For matter to move, force is required. Light is always moving. Once again, pure classical mechanics.

For a long, long time, Newton's particle theory was the accepted definition of light. There's a certain amount of irony in this, however—since Newton was wrong. Or, to be more precise, he just didn't have the whole picture. In other words, he wasn't completely correct. Purists will want to make the point that this is different from "wrong." However, just as no one wanted to disagree with Aristotle for 2,000 years, no one wanted to mess with Newton, either. After all, who was going to be foolish enough to argue with him? He was the guy, after all, who came up with all the laws that straightened out 2,000 years' worth of physics history.

Then in 1800, a guy named Thomas Young caused a giant fuss when he began to insist that light was a wave. And lots of people agreed with him.

The funny thing was that he was partially wrong, too.

The Secret Life of Light

The faint of heart may find the following section a little disturbing. For one thing, it will cause you to consider that everything that you have read in this book thus far may be wrong. Even more upsetting than that, it may cause you to consider that everything you know about the entire physical universe may also be wrong.

So, here's where we stood: Newton said light is a particle. And he was right in the sense that light obeys many of the Newtonian laws of force and inertia as they have applied to other matter of a particle nature . . . like an apple falling from a tree. That is, light follows those laws until you look a little closer.

Young said light is a wave. And he was right in the sense that light, like sound, can undergo interference when two waves meet. Particles don't experience interference when they meet; they just have a collision and share a transfer of energy. Therefore, light must be a wave. And this view of light stuck for most of the rest of the century.

The problem with the scenario above is that sometimes light doesn't act like a wave; sometimes it acts like a particle. And sometimes it doesn't act like a particle, but it acts like a wave. When it acts like one or the other we have no problem, since waves and particles both act according to Newton's laws. On the other hand, when light acts like a "wavicle" (that is, sometimes it acts like a wave and then turns around and immediately begins to act like a particle . . . and then back again faster than Bill Clinton can act like a man with presidential and moral character and then walk into the Oval Office, close the door, and carry on in a completely dishonorable manner), it does things that just don't fit with our expectations of how things should behave, given our understanding of the universe on the basis of Newton's laws.

And all of this, as you can imagine, is the sign of an even bigger problem. Physics is a comprehensive science. Things are either true . . . or they're not. And Newton's laws, the basis for all the exciting discoveries about mechanics and electricity and magnetism and heat and sound we've been discussing for the past several pages, were beginning to look incorrect. It was, you might say . . .

. . . THE END OF CLASSICAL PHYSICS

The following may feel like a digression. Don't worry, it's not.

WHAT'S AN ATOM?

In the late 1800s, physicists began theorizing that all matter was made up of tiny little particles that they called atoms. But there was a problem with this theory: The only evidence anyone had of atoms' existence was purely circumstantial.

For instance, it was believed that the way in which gas behaved under pressure proved the existence of the atom. After all, gas will fill any container you put it in, but the pressure on the walls of the container will increase if you reduce the volume of the container. How else do you explain this if you don't use a theory that involves tiny bits of matter like atoms that are always moving, jostling, and bumping into each other and bumping into the walls of the container? Take away half the space these tiny bits have to roam in and suddenly, you get twice as much pressure.

There was still one problem with this theory, however; although in theory, the existence of atoms explained a lot, no one had ever actually seen an atom . . .

. . . until 1897, when physicist Joseph John Thomson announced the discovery of what he called the electron. Eventually, this so impressed everyone that he was knighted—that's Sir Thomson to you.

Boldly and bravely, he proclaimed that this tiny but fast little particle with a negative charge, which he had discovered while playing around with cathode ray tubes, was not an atom but a particle that existed inside the atom.

On the basis of what he had learned, Thomson went on to theorize that the atom itself might be a sphere with his electrons floating around inside. He called his picture of the atom the "plum pudding" model because he envisioned the electrons floating around inside the atom like some plums you might find in a bowl of plum pudding.

How significant and startling was Thomson's discovery? Years later he confessed to the following exchange after he first made his discovery public: "I was told long afterward by a distinguished physicist who had been present at my lecture that he thought I had been pulling their leg."

ERNEST RUTHERFORD DISCOVERS THE NUCLEUS

By bombarding Thomson's atoms with alpha particles, Ernest Rutherford, first Baron Rutherford of Nelson and Cambridge, discovered the nucleus. Alpha particles, along with beta and gamma rays, are the three main components of radiation—and Rutherford identified them, too. Anyway, he quickly noted that most alpha particles passed right through the atom as if nothing were there—yet a few of the alpha particles, perhaps 1 in 1,000, would be deflected by something to a slight degree and even fewer, perhaps 1 in 10,000, would be greatly deflected to such a degree that it almost appeared as if they had been shot back at the observer.

> It was quite the most incredible event that has ever happened to me in my life. It was almost as incredible as if you fired a 15-inch shell at a piece of tissue paper and it came back and hit you.
>
> —Ernest Rutherford, reflecting on the first time he saw an alpha particle deflect

Rutherford reasoned that he could use this information to "paint" a picture of the inside of an atom, which he saw as having a nucleus in the center, surrounded by rings of electrons, reminiscent of our very own solar system.

In fact, by plugging his data into the same formulas relating to electrical fields, mass, and acceleration that Thomson had used, Rutherford found he could actually prove that Thomson's plum pudding model was wrong. After all, if Thomson were correct, then Rutherford's alpha particles should strike (or at least be deflected by the negative charge of the electrons) far more often than the experiment showed.

By contrast, Rutherford's experiment showed that most of the mass of the atom was concentrated in the tiny nucleus at the center of the atom—not spread out over the entire area like pudding. In fact, the Rutherford deflection was caused entirely by the nucleus, and it demonstrated that the atom was mostly empty space, in contrast to the plum pudding model. Instead, Rutherford hypothesized that there could only be one explanation for why the alpha particles would behave as they did: The inside of the atom must be mostly empty space, allowing all but 1 in 1,000 of the alpha particles to pass through undeflected.

On the basis of this theory, Rutherford was able to predict the size of the area within the atom that contained matter, which he called the nucleus. His colleagues were so impressed that they named the rutherford, a unit of radioactivity, after him.

This digression—which was not really a digression—ends here.

THE BIRTH OF QUANTUM THEORY

In 1900, a guy named Max Planck cooked up a theory about how energy in an atom is different than energy in Newton's world of mechanics. In mechanics, anything that moves and is made of matter carries energy. So even if you're debating whether light is a wave or light is a particle, from the perspective of the transmission of energy, it doesn't matter. Both waves and particles, by their very nature, are moving and are therefore carrying energy at all times. It's as simple as that.

It would be nice if matter inside the atom played by the same rules, but it doesn't. Or so said Planck. Inside the atom, he said, matter is always moving, but it is not always radiating energy.

AND LET'S NOT FORGET THE ETHER QUESTION

Oh. You didn't know there was an ether question?

Well, it goes like this: If Thomas Young was right that light moves as a wave, then as a wave, it had to move through something, right? After all, without spectators, you've got no wave at a ball game, and without water, you've got no waves at the beach. Therefore, for light to be a wave, it had to move through—ta da!—the ether.

Well, looking for the ether is sort of like looking for the emperor's new clothes . . . it's invisible and therefore pretty hard to prove or disprove.

Unless you're in Cleveland.

That was the home of two college professors named Albert Michelson and Edward Morley in 1887. They figured that if they could measure the speed of the Earth relative to the speed of the ether (which wasn't moving at all—remember the duck in the tub?), they would have solid proof that the ether exists.

For nonscientists, this may sound like a crazy idea. An impossible idea. And you are right, for the Michelson–Morley experiment never proved a darn thing except that the proposed ether was only a lot of hot air.

"Well, why not," we ask as Newton again turns over in his grave.

According to Planck's theory, light is sort of like a vending machine; if you've got enough money, you can buy a soda. If you don't, you're out of luck. Of course, at any vending machine, the price for different items can vary, and the same was true with Planck's theory. The "cost" of the energy required to "buy" infrared light was lower than the cost of green light, which was lower than the "cost" of ultraviolet light, for example. In every case, however, a minimum "charge" was involved. And if the energy fell short of this minimum energy—which he called a quantum—then the deal was off. There would be no light.

Although Planck was the one to propose the existence of these little packets of energy, called light quanta, or photons, it was Albert Einstein who proved it was true. In fact, it was for this that Einstein won the Nobel Prize . . . not for either of his theories of relativity.

Time for another digression-that-is-not.

THE ONE BOOK I WOULD READ . . .

My first real exposure to quantum mechanics came in 1958. I was then a graduate student at the University of California at Los Angeles, pursuing an advanced degree in physics. One of the textbooks required for our course in quantum mechanics was David Bohm's *Quantum Theory*. It is an unusual college physics textbook. As you may have discovered, physics textbooks are normally quite dry and loaded with seemingly undecipherable formulae apparently created by machines rather than people. Bohm's book was an exception. It had more words than formulae. It dealt with questions concerning topics that were apparently unrelated to physics. "The Indivisible Unity of the World," "The Need for a Nonmechanical Description" of nature, "The Uncertainty Principle and Certain Aspects of Our Thought Processes," and "The Paradox of Einstein, Rosen, and Podolsky" were some of the topics that Bohm considered and that were to have a great influence on my own thinking.

— Fred Alan Wolf, Ph.D., author of *Taking the Quantum Leap: The New Physics for Nonscientists*

ALBERT EINSTEIN MOUNTS A SIDE ATTACK ON NEWTON AND CLASSICAL PHYSICS

While most everyone else was arguing about the structure of the atom and how the atom worked (and if there was such a thing as an atom at all), Einstein went spinning off in an entirely different direction. His goal was to isolate and prove the existence of ether— or to throw the whole darn theory away.

LOVE YOUR UNDERACHIEVER

Poor Mr. and Mrs. Einstein. By anyone's standards, their son Albert was a loser:

- He didn't start talking until he was three.
- He didn't like school, so he quit going at the age of 15.
- When he decided to take the entrance exam to the Swiss National Polytechnic in Zürich instead of completing high school, he didn't pass the test.
- After returning to high school and getting a diploma, he was finally admitted to college, but he often cut class.
- He did finally graduate from college—but only because a good friend lent him his class notes.

- After completing college, Einstein was qualified to teach math and physics, but he couldn't get a job.
- Although he did work as a teacher on a temporary basis for a while, he finally gave up hope that he would ever get a full-time job, so he went to work as a clerk in the Swiss patent office instead. This is where he remained for seven years.

OR DID ALBERT HAVE A MASTER PLAN?

- Although Einstein started talking late, once he started talking he spoke in complete sentences.
- Although he hated school, he didn't hate learning. In fact, at age 12 he taught himself Euclidean geometry and in the same year found a completely new proof for the Pythagorean theorem.
- Although he finally gave up hope that he would ever find a permanent teaching job and took a job at the Swiss patent office instead, this turned out to be a blessing. As he once said, his job at the patent office was such that he could leave at the end of the day free to go home at night and work on his science, whereas he saw that an academic career ". . . puts a young man into a kind of embarrassing position by requiring him to produce scientific publications in impressive quantity—a seduction into superficiality . . ." This was the same superficiality that had caused Einstein to cut so many of his high school and college classes, anyway.

EINSTEIN HAS A GOOD YEAR

He introduced his theory of special relativity in June 1905. But that's not all he did that year:

- March 1905, he developed the quantum theory of light, a theory that was equally profound and shocking at the time.
- In April and May, he published two articles that together proved the existence of the atom.
- In June, he presented his article on special relativity. Ever nimble-minded, he treated light as a wave in this article but treated it as a particle in the article on quantum theory that he had presented earlier, in March.

- Later in the year he made an addition to his theory of special relativity. He added the now-famous formula $E = mc^2$. Is it any wonder that the year 1905 is called Einstein's *annus mirabilis,* or miracle year?

AL HAS AN EVEN BETTER YEAR . . . WHILE TAKING A ROAD TRIP 42 YEARS AFTER HE DIED

As part of the process of writing an article for *Harper's* magazine, writer Michael Paterniti drove from New Jersey to California in a Buick Skylark with an 84-year-old pathologist and a Tupperware bowl full of Albert Einstein's brain.

We kid you not.

It turns out that the pathologist, Thomas Harvey, was the guy who did Einstein's autopsy back in 1955—and when he was done, he took Einstein's brain home with him.

Forty-two years later, Harvey drove the brain to California to show it to Einstein's granddaughter.

Of course, this was a road trip in every classic sense of the word, so Harvey wanted to show Einstein a good time. Therefore, he made a stop in Lawrence, Kansas, so he could introduce one cultural icon—Einstein—to another, Harvey's longtime friend and Beat Generation poet William Burroughs.

And the writer—Paterniti—and *Harper's* magazine swear every word of this is true.

$E = mc^2$*

Einstein's discovery of relativity was sort of like Columbus's discovery of America. Columbus was looking for India, so he landed at a place that looked like what he thought India should look like and he called the people he met there Indians.

Einstein was looking for relativity, so that's what he called what he found. Only 10 years later did he realize that what he had originally discovered was a special case of a more general law that was still out there waiting to be found. Today, Einstein's 1905 theory is known as special relativity and his 1915 theory is known as general relativity.

---

* $E = (\text{mass} \times \text{confusion})^2$.

> **HOW ALBERT EINSTEIN TRICKED THE TEACHER**
>
> Four years after he introduced his theory of general relativity, Einstein stopped off at his old school to pay a visit. The headmaster, remembering what a poor student Einstein had been, quickly jumped to the conclusion that his former student had returned to campus to beg for money or food.

"ISN'T THAT SPECIAL?" OR, HOW TO TELL SPECIAL RELATIVITY FROM THE OTHER, MORE GENERAL KIND

The easiest way to understand what Einstein had in mind is to go back to one of Galileo's main precepts in his concept of inertia: that it is impossible to tell the difference between a body at rest and a body that is moving at a uniform rate of motion—because applying the laws of classical physics in either situation will always bear the same results.

Just as a review, let's test this out. Hold this book in your hand and then drop it. It falls in your lap. Now drive your car up on the highway, lock your cruise control in at 65 miles per hour, and drop the book again. Does it fly back at you and hit you in the chest or does it fall into your lap again?

The conclusion we can draw from this little experiment is that whether you are standing still or moving really fast at a uniform rate, the same basic laws of classical physics apply and will produce the same results.

"WHY DIDN'T I THINK OF THAT?"

Let's say scientific discoveries come in two flavors: "Duh!" and "Huh?"

When a discovery of the "Duh!" variety is made, everyone in that associated field slaps his or her forehead and mumbles, "Well, there goes the Nobel Prize! Why didn't I think of that?" A "Duh!" discovery is so obvious, anyone (well, almost anyone) could have found it if he or she had merely looked in the right place or looked in the same old places in a slightly different way. Discoveries like these are made not just in science but by people in all walks of life, all the time—like when you spend an hour looking for your keys only to discover that you had walked past them numerous times as you searched for them.

"Duh!" you say, then slap your forehead and move on.

Galileo's gravity, in a sense, was a "Duh!" Obviously, something was pulling everything down toward the ground. Someone just had to figure out what that something was.

On the other hand, Newton's gravity was a "Huh?" No matter where you looked or how you looked at what was in front of you, it took some kind of weird and twisted thought (or divine inspiration), beyond the circumstances and evidence in front of your face, for you to figure it out.

One of the interesting ways in which science and history interact is this: Something that is a "Huh?" for one generation (i.e. Newton's) will become a "Duh!" for succeeding generations. For instance, once everyone accepts Newton's explanation for gravity as a given, we expect the rest of the world to act accordingly—but the first people to readily accept the truth of a huge scientific discovery like gravity are those who weren't alive before the discovery was made.

Thus, the study of classical physics and mechanics was turbocharged once Newton's explanation of gravity passed onto a new generation of scientists and gravity became part of accepted thought. People naturally applied it to other areas of life and the universe as a matter of course, and the more they expected to see things in the universe act according to the laws of gravity and the more the universe did as they expected, the easier it became to make discoveries. "Duh!"

A perfect example of how we the people assimilate scientific discoveries in succeeding generations is to recall how controversial it was for Galileo to assert that the Earth was moving. For his relentless insistence that this was true, he was put under house arrest for the last 10 years of his life. (Of course, it could have been worse. Just ask Giordano Bruno [1548–1600], who lived around the same time as Galileo [1564–1642] but was burned at the stake for throwing his support to Copernicus and the sun-based theory of the universe).

But let's get back to Galileo. Recall how he proved his assertion: He climbed the mast of a moving ship and dropped a ball. And the ball acted just as it would have acted had he dropped the ball from atop a pole that was anchored into solid ground . . . it fell straight down. Today, Galileo's experiment is a "gimme"—after all, we "know" the Earth is moving because we've sent people up into space to witness it. Galileo, on the other hand, had to make his assertion without the benefit of any eyewitnesses.

The ultimate "Huh?" discovery, of course, is Einstein's theory of relativity. It messes with time, something we all think we understand . . . until we try to understand what Einstein was trying to say.

FROM "HOW COME . . . ?" TO "WHAT IF . . . ?"

We could also say that Einstein took physics from an era of asking questions about observable phenomena (How come things fall? How come magnets attract? How come light behaves as it does?) to an era of asking questions about elements of nature that are unseen and that can only be dreamed about (What if I was moving at the speed of light and I turned on a light?).

Here's why this is significant: Einstein wanted to find a way to demonstrate that Galileo's inertia was right not only for classical physics but also for modern physics. (But be careful here. This is not the same thing as saying that the laws of classical physics and modern theory are the same. They have many differences—but inertia, Einstein believed, wasn't one of them.)

NOW IT'S TIME FOR AN EINSTEIN ANECDOTE

When he was a little boy, Einstein once asked his teacher what would happen if he were traveling at the speed of light and then turned on a light. Ponder this for a moment and you will be ready to get down to the essence of special relativity, which Einstein was able to state in two postulates:

1. The speed of light is the same for all observers, regardless of their motion relative to the source of light. To illustrate, when I enter a room and flip on the light, the light illuminates the room and radiates away from me in all directions at the speed of light. This much, Einstein understood as a young boy, but what he wanted to know was what he would see if he were moving at the speed of light as he entered this room and continued moving at the speed of light as he flipped on the switch. The answer lies in the next postulate:

2. Two observers moving at constant speed (but not necessarily the *same* constant speed) will observe the same results from all physical laws.

"Huh?"

For postulates 1 and 2 to be true, there can be only one answer to Einstein's question: if he is traveling at the speed of light and you flip on a light while standing still, you and he both see the same thing—a light moving at 186,000 miles per second—even though you are standing still and he is already traveling at 186,000 miles per second. What we discover is that the two laws are the same but the observations are different.

Here is why relativity is so hard to understand: According to classical physics, if you're driving in a car at 70 miles per hour and you throw a ball at 1 mile per hour at the windshield, the ball will appear to be moving at 1 mile per hour to you but would appear to be moving at 71 miles per hour (the 70 miles per hour of the car plus the 1 mile per hour of the ball) to an observer standing at the side of the road. But according to Einstein's theory, light does not behave as the ball does—and that's where relativity departs from the everyday reality of folks like us.

To make this easy, let's slow the speed of light down to 140 miles per hour.

Then, let's imagine that we are driving at half this new speed of light, or 70 miles per hour.

And, finally, let's imagine you throw a ball at the windshield at the speed of 1 mile per hour.

To you, the ball moves at 1 mile per hour. To an observer on the side of the road, the ball moves at 71 miles per hour.

Next, as you drive at 70 miles per hour, or half our new speed of light, you turn on a flashlight and aim it out the windshield.

Given our scenario with the ball, we would expect to see the light travel away from us at 140 miles per hour, our new speed of light. Meanwhile, we would expect an observer on the side of the road to see the light travel away from him at 210 miles per hour, or the speed of the car, 70 miles per hour, plus our reduced speed of light, 140 miles per hour.

But what Einstein would tell us really happens is that both we in the car and the observer outside see the same thing . . . the light traveling away from us at the same speed, 140 miles per hour, our speed of light.

Don't worry if this last sentence makes your head hurt. It should. And the implications of Einstein's theory might even make your stomach churn a bit the way it does as you come to a peak on a roller coaster and are about to go tumbling downward.

Meanwhile, it's worth taking a moment to talk about the speed of light. At 186,000 miles per second, not only is it a good way to measure how fast something is moving but it is also a good way to measure time. For instance, imagine you are at work and your car has broken down. You need a ride home, so you ask a coworker for a lift. "It's not far," you say. "It's only about ten minutes away." You could say it's only a couple of miles away, but given all the stop-lights and traffic between here and there, you know that citing the distance from your house to the office doesn't really give an accurate picture of what's involved, so you describe the distance (or space) from point A to point B using time as your frame of reference. When we are talking about relativity, we are talking about space and time in the same way.

"WHAT DID YOU SEE AND WHEN DID YOU SEE IT?"

Albert Einstein was a simple guy. He kept his pants up with a piece of rope and often forgot to comb his hair. He was also known to be pacifistic and gentle, but his theory of relativity is full of conflict over who saw what and when he saw it.

He was a big fan of what he called *Gedanken,* or hypothetical thought experiments. When he wanted to figure out how something worked that involved elements of nature he couldn't observe directly, he would imagine the reality of the experiment instead. Baseball fans have been doing this for years . . . sort of like: "Who do you think would have won the 1953 World Series if Ed Lopat had been pitching for the Brooklyn Dodgers instead?"

But let's put all questions about the New York Yankees aside for a few moments and think about relativity. In this *Gedanken,* we're going to again work with the premise that we are traveling in a car at 70 miles per hour and that the speed of light is 140 miles per hour. Given that premise, we have some unusual scenarios to consider.

FASTER THAN A SPEEDING GEDANKEN . . .

SCENARIO 1	SCENARIO 2
Who sees what:	**Who sees what:**
■ You're in your car. Again, you throw the ball at the windshield. Again, you see the ball move at 1 mile per hour. ■ On the other hand, let's say you're standing on the street as the same car drives past. Just as the car crosses your path, you see the passenger in the car throw the ball at the windshield. You see the ball move at the speed of the car, 70 miles per hour, plus the speed of the ball, 1 mile per hour—or a total of 71 miles per hour.	■ You're in the car again. You shine your flashlight at the windshield. What you see is the beam of light zipping away from you at 140 miles per hour, our reduced speed of light. ■ You're standing on the street again. The same car passes you, and as it goes by, the passenger turns on a flashlight and points it out the windshield. Given what you know about classical physics and speed and velocity, you expect to see the light zip away at 140 miles per hour, our reduced speed of light, plus 70 miles per hour, the speed of the car.
Conclusion:	**Conclusion:**
There's nothing new going on here; the difference in speed is attributable to the velocity of the car. When you're in the car moving with it, its velocity is not a factor in what you see. When you're on the outside, your observation is a combination of the velocity of the car plus the velocity of the ball.	Ah ha! You might think the person in the car and the person on the street would see two different things—but you'd be wrong. Light is not governed by the same mechanical laws that govern matter. When you understand that, you will have gotten to the heart of Albert Einstein's theory: Whether you are in the car moving at 70 miles per hour or standing on the street corner, the speed of a light moving away from you is the same. Unlike scenario 1, in which the observer in the car sees something completely different than what the observer on the street sees, in scenario 2 both observers see exactly the same thing.

THE BOTTOM LINE ON EINSTEIN'S SPECIAL RELATIVITY: MOTION AFFECTS MEASUREMENT

In relativity, it's all about your frame of reference. Whether you are talking about time or space or speed or distance, it's not the specific quantities we're concerned with but the quantity of your time or space or speed or distance relative to another body.

"THIS IS EINSTEIN TO MISSION CONTROL"

Forget what you know. The universe operates on a different set of rules than you think it does.

Relativity Rule #1

Time is not a constant . . . the faster you go, the slower it goes. And vice versa.

Here's an example: you buy a new Rolex watch, drive to the airport, and just before you jump on a plane that will fly you around the world, you synchronize it with the new Rolex you just bought for your mate. "Let's keep an eye on these things and make sure they really work," you both agree, and then off you go. A week later, you return from your trip around the world, your mate meets you at the airport, and you compare the time on your watches. Yours is running hours slower, even though you spent countless hours watching every second tick off with perfect precision. Furious, you storm back to the Rolex dealer, ready to raise the roof. As you get halfway through your story, the salesman smiles and pulls on one corner of his mustache, then holds up his hand to stop you.

"But sir," he says, "I can assure you there is nothing wrong with the watch. Because it is a Rolex, I am sure it keeps perfect time. If you want to finger a culprit, I think you'll find that Albert Einstein's theory of special relativity is to blame."

And as frustrated and embarrassed as you might be, your Rolex salesman would be right. The secret here is that to you, your watch was at rest while you were traveling—whereas to your mate, it was in motion. Again, this boils down to the essence of special relativity: Motion affects measurement.

Still struggling with this? Let's look at it another way.

Let's say you hire a friend to videotape your wedding. "But I can only stay an hour," he says, so you give him a 60-minute tape. He decides that he will call the video "One Last Hour of Bachelorhood," so he begins the videotape with a shot of your watch, periodically takes footage of the moving hands of the watch throughout the ceremony, and ends the video with the watch showing that an hour has passed.

Later, when showing the video to friends, you decide to fast-forward through the first 15 minutes of the tape and get to the kiss that consummates the marriage. As the tape begins, you see your watch on the screen; it is 12:00. You then fast-forward, and less than three minutes later, your watch on the video shows that it is 12:15. Thus, using the modern technology of a VCR, you have demonstrated how time can move at two different speeds. (We don't want you to think the scenario above is an actual example of relativity; however, instead it is an analogy that makes relativity easier to comprehend.)

Relativity Rule #2
Moving objects appear to be shortened in the direction in which they are moving.
Of course.

After all, if time is going to change as you move faster, doesn't it also make sense that the objects you pass as you move through that faster time will change as well?

This phenomenon is more common than you might think. To give you a flavor of it, let's say you normally drive to work at 50 miles per hour and it takes you 10 minutes to get there. One day, however, you oversleep and are late for an important meeting, so you jump in your car and rocket off toward your office with the pedal to the metal at 100 miles per hour. Along the way, if you took the time to look, you would notice that passing 10 minutes of scenery in five minutes has an interesting impact on it . . . since you have half the time to drive by the local barbershop, for instance, it appears to be somewhere near half its usual size. (Interestingly, if you had awakened early and had walked to work instead of driving, then the barbershop would appear to be much bigger to you than it usually does as you pass it at 50 miles per hour).

Although it is the buildings that appear to be different when you drive at 100 miles per hour, what has really changed is time. After all, if you were standing in front of the barbershop while one car drove past at 100 miles per hour and another car drove past at 50 miles per hour, you and the barbershop would look different to the two drivers but you wouldn't feel any different.

Relativity Rule #3
There is no such thing as simultaneous events.
On one hand, this seems to be a contradiction. You look up at the sky and see an exploding star. "Look," you say, "I saw it just as it happened." A textbook example of a simultaneous event if we've ever heard of one!

But it's a textbook example of simultaneity only because it so perfectly demonstrates why there is no such thing. After all, even though you were lucky enough to see the star "just as it exploded," the truth is that this star actually exploded several thousand years ago and is long gone by now, and the only reason you could see it exploding now is that it took this long for the star's final light waves to travel the distance from space to Earth where your lucky eyes just happened to be looking at the right time to see it.

Given the great distance between you and the star, it is easy to see that what looked like a simultaneous event was actually separated by thousands of years. It gets a little harder to argue against simultaneity when two events occur with only a small amount of distance between them. However, regardless of how close two observers are to one another, there is always a time difference between them and, therefore, there can never be simultaneous events.

For instance, when the the town fire station's alarm sounds to mark noon, we assume that everyone hears it simultaneously and knocks off for lunch at the same time.

But if you want to see how wrong that notion is, try this little experiment: get together with a friend and two cellular phones. Drop your friend off with one phone about a mile from the fire station and then drive on another mile so there are two miles between you and the fire station, with your friend halfway between.

At 11:55, you should use your cellular phone to call your friend. Since you have a few minutes, go ahead and catch up on gossip—

but when the sound of the fire alarm interrupts, you're going to be surprised by what you hear . . . first, you will hear the alarm over the phone as it blares where your friend is, and then moments later the alarm will blare again as the sound reaches you.

So much for simultaneous events. No matter how far away from or close to someone else you are, when an event happens to you it cannot also be happening to the other person. There must be some differential in time.

Relativity Rule #4
The speed of light is the fastest rate at which any object may travel.

Here's the easiest rule of relativity you'll ever find: Nothing can travel faster than the speed of light. At 186,000 miles per second, it is nature's fundamental speed limit.

As you can see, light does some very strange things. Sometimes it seems to travel very fast and sometimes it seems to travel very slow—but the truth is, it's always traveling at the same speed regardless of how fast you are traveling relative to it.

THE ONE BOOK I WOULD READ . . .

The book by Karl Darrow, *Introduction to Contemporary Physics,* which I read when I was first graduating from college. I was getting a little bit discouraged about physics because it seemed to be getting mathematical. But this book told about the experimental basis for a lot of the discoveries of the twentieth century instead of only the mathematical side. It really was fascinating.

—Arthur Schawlow, winner of the 1981 Nobel Prize in physics

For a great survey of the history of physics, see *Physics for Poets,* by Robert H. March.

SPACE AND TIME: IT'S ALL RELATIVE . . . ISN'T IT?

A favorite paradox for many physicists involves a set of twins. One twin boards a spaceship that is capable of near light-year speed travel and he sets off for a star that is 20 light-years away; the other twin stays at home.

MY TIME ISN'T YOUR TIME

Mathematician Hermann Minkowski Invents a New Word: *Space–time*	Traveling at the Speed of Light: Can I Get Frequent Flyer Miles for That?	Is It a Light Year, a Year of Light, or Just a Long, Long Way to Go?
Imagine, again, you have asked a friend for a ride home from work. Again, you say that it will only take 10 minutes to get there, which is all very nice when you're moving at 50 miles per hour. But when you're moving at the speed of light, 186,000 miles per second, you can't just say that it'll take 10 minutes to get from here to there, because for one thing, when you're moving at that speed, by the time you finish saying the sentence "It'll take just ten minutes" you will have already traveled the distance from here to the moon (and back, if you talk with a southern drawl.)	Think about it: if the speed of light is 186,000 miles per second, then it would take about eight minutes to get from here to the sun—but it would only take a second and a half to get to the moon and a little more than two minutes to get from here to Mars.	As you might guess, a light-year is how far light can travel in one of our years; close to 6 trillion miles. (That's the number of seconds in a year: 60 seconds × 60 minutes × 24 hours × 365 days = 31,536,000 seconds. So, 31,536,000 seconds × 186,000 miles per second = 5,865,696,000,000 miles per year.)

Think about it: if the speed of light is 186,000 miles per second, then it would take about eight minutes to get from here to the sun—but it would only take a second and a half to get to the moon and a little more than two minutes to get from here to Mars.

So let's try to figure this out: If we plan a trip to the moon in January, how many free trips to Disney World and Epcot Center will that give us if the standard airline "fee" for a round-trip ticket within the continental United States is 25,000 miles?

Hmm, let's see:

1. If the speed of light is 186,000 miles per second and it takes a second and a half to get to the moon at the speed of light,

As you might guess, a light-year is how far light can travel in one of our years; close to 6 trillion miles. (That's the number of seconds in a year: 60 seconds × 60 minutes × 24 hours × 365 days = 31,536,000 seconds. So, 31,536,000 seconds × 186,000 miles per second = 5,865,696,000,000 miles per year.)

On the other hand, you might want to know how long one of our years—31,536,000 seconds—would last if you were traveling at the speed of light. The answer: almost three minutes (31,536,000 seconds ÷ 186,000 miles per second = 169.54 seconds = 169.54 ÷ 60 [seconds in a minute] = 2.83 minutes).

It might stand to reason that one of our years—31,536,000 seconds—would last about three minutes if you were traveling at the speed of light. However, as with many things in physics, the math does not equal the reality.

And for inquiring minds, a

Imagine, again, you have asked a friend for a ride home from work. Again, you say that it will only take 10 minutes to get there, which is all very nice when you're moving at 50 miles per hour. But when you're moving at the speed of light, 186,000 miles per second, you can't just say that it'll take 10 minutes to get from here to there, because for one thing, when you're moving at that speed, by the time you finish saying the sentence "It'll take just ten minutes" you will have already traveled the distance from here to the moon (and back, if you talk with a southern drawl.)

Meanwhile, as friend A drives you home, friend B pulls alongside. You wave to him, indicating that he should pull in behind and follow you and friend A to your house.

When friend B follows friend A at the exact same speed, then both

Mathematician Hermann Minkowski Invents a New Word: *Space–time*	Traveling at the Speed of Light: Can I Get Frequent Flyer Miles for That?	Is It a Light Year, a Year of Light, or Just a Long, Long Way to Go?

will be in the same space–time continuum. After all, it could be said that from B's perspective, A is not moving, since the space between the two cars never changes. But if B stops at a light while A races through it, the space–time continuum of the two cars changes—and if A is traveling near the speed of light as she races away from B, then the only way A can describe her position relative to B (and the only way B can describe *her* position relative to A) is by using coordinates of space as well as the coordinate of time.

Now, let's use another familiar analogy—your wedding video.

You can again see space–time in action by playing your wedding video for friends three months, three years, and 30 years after the wedding. Two questions will always arise:

▪ Where was your wedding? (Which you can answer using the

then the moon must be 279,000 miles from Earth: 186,000 + (186,000 ÷ 2) = 186,000 + 93,000 = 279,000.

2. Of course, we're going to travel to the moon and back, so our total mileage for this trip would be 558,000 frequent flyer miles: 279,000 + 279,000 = 558,000.

3. At the time of this writing, a round-trip ticket to Disneyworld "costs" 25,000 miles.

4. Therefore, we would have enough miles for 22.32 round-trip tickets: 558,000 ÷ 25,000 = 22.32.

Our conclusion: use those miles quick. You can be sure that once people start traveling to the moon on a regular basis, the number of miles you need to get a round-trip ticket for a trip within the continental United

long, long way to go is a light-year in dog years— 41,059,872,000,000 miles: 5,865,696,000,000 × 7 years = 41,059,872,000,000 miles.

Mathematician Hermann Minkowski Invents a New Word: *Space–time*	Traveling at the Speed of Light: Can I Get Frequent Flyer Miles for That?	Is It a Light Year, a Year of Light, or Just a Long, Long Way to Go?
three-coordinate system of space) ▪ When was your wedding? (Which has one coordinate—time) Just as you would never try to determine the size of a room without measuring all three coordinates (height, width, length), it is equally impossible to talk about an event without the same coordinates of space as well as one additional coordinate—time.	States is sure to go up. A lot.	

As you might expect, it takes the traveling twin 20 years to get to the star, a few minutes to have lunch, and another 20 years to get home.

When he lands back on Earth, he is greeted by a guy who could easily be his brother if only he weren't so darn old.

The funny thing is, as the brother who stayed home approaches the spaceship to greet the brother who took the trip, he is shocked that a guy who looked just like his brother did 40 years ago is climbing down out of the hatch.

To explain what has happened, let's imagine there is a flashing light is mounted on a tower on Earth that flashes every five minutes, or 12 times an hour. But that's true only when you are on Earth. If you are on a rocket, you won't see a flash every five minutes as you fly away from the Earth; you will see them less frequently because as you get farther away, it takes the light longer to reach you. The same should also be true on your way back home— you should see the signals more frequently because you are getting closer and closer to the Earth.

THE EINSTEIN ADDENDUM: GENERAL RELATIVITY

In 1907, two years after he published his first paper on special relativity, Einstein was suddenly struck by what he later called the happiest thought of his life: "I was sitting in a chair in the patent office in Bern when all of sudden a thought occurred to me. 'If a person falls freely he will not feel his own weight.' I was startled. This simple thought made a deep impression on me. It impelled me toward a theory of gravitation."

So what was wrong with *Newton's* law of gravitation?

Nothing. Except it wasn't quite correct. It couldn't be.

After all, Newton himself said everything attracts everything else with a force that is proportional to their two masses and inversely proportional to the square of the distance between them—or $(m)(m') \div d^2$—but Einstein's special relativity had redefined mass, so how could Newton's formula still be right?

To illustrate, we flip a coin:

HEADS OR TAILS?	
When Isaac Newton Flips the Coin	When Albert Einstein Flips the Coin
The ground and the coin experience the mutual attraction of gravity. The coin hits the ground at Newton's feet in less than half a second.	What happens depends on whether you are Einstein or the coin, for as Einstein said, "If a person falls freely he will not feel his own weight."
	However, from Einstein's perspective, the coin behaves exactly the same way it does for Newton—as if gravity were pulling it to the ground. Compare that with the coin, which has an experience of gravitylessness as it falls—and if the coin were you, and we blindfolded you, put you in a box so you couldn't feel the wind, and flipped you right off the Empire State Building, you'd feel like you were floating . . . until you hit the ground.

Einstein and Newton flip again. This time, they are in a rocket with uniform speed, far away from any large bodies that would have a gravitational influence.

When Isaac Newton Flips the Coin	When Albert Einstein Flips the Coin
The coin floats in the air. As does Newton, the flipper.	For the coin, the experience is the same as above. Einstein, the flipper, floats in the air, too.

Now they flip again, but this time they are out in space on an *accelerating* rocket ship, far away from any large bodies that would have a gravitational influence.

When Newton flips the coin, it clatters to the floor, just as on Earth.	Again, what happens depends on whether you are Einstein or the coin. But here, Einstein understands why the result is exactly the same as if he were standing on the ground, because it is impossible to tell the difference between gravitation and acceleration. The relevance of this discovery will become apparent shortly.

"WHAT'S IT ALL ABOUT, ALFIE?"

For Newton, mass was a constant. On the other hand, Einstein's theory of gravitation (a.k.a. general relativity) takes into account the fact that mass will change according to (1) speed and (2) who's looking and from where.

For most of us, Newton's gravity provides a picture that is so close to reality that there's no sense arguing about it. On the other hand, when you start dropping things from hundreds of feet or traveling in spaceships, Einstein's explanation is the right one to use.

Was Newton wrong? No, just half right. His theory of gravity predicted how the coin would behave from the perspective of the flipper. But as Einstein taught us, that's only half the story. Therefore, his theory, general relativity, accounts for the behavior of the coin regardless of who is looking and where they are looking from.

Einstein's theory of general relativity causes us to alter our view of some other aspects of the physical world as well:

YIKES! LIGHT BENDS!

And it can all be proven using Einstein's theory of general relativity.

Proving Space Is Warped	Proving Light Bends
Think of the globe as space. Start somewhere around the equator and use your fingers to trace the route of two adjacent longitudinal lines (the ones that go up and down) toward the North Pole—and you discover very quickly that the idea that parallel lines never insect is a myth. It's true only when you're dealing with a two-dimensional universe. Space, like your globe, is curved. Therefore, the shortest distance between two points is often not a straight line . . . and that's also why when you drop a ball from waist height it falls on a curve and why light bends, as well.	We're in a rocket ship. And we've got a another rubber ball with us. If we throw the ball while the ship is standing still or moving at a uniform rate of speed, both an observer on the ship and another observer outside the ship will see the ball fly in a straight line across the ship and hit the wall at (about) the same height from which it was thrown. However, if we throw the ball across the ship while the ship is accelerating upward, the observers see two different things.

Coming to a complete understanding of Einstein's theories of special and general relativity is a daunting task, even for scientists. One place to start is at the source, with Einstein's own *Relativity: The Special and General Theory*. But don't be discouraged if Einstein's

SPACE IS WARPED!

OUTSIDE OBSERVER

INSIDE OBSERVER

The outside observer sees that the ball still travels in a straight line but hits the opposite wall below the height from which it was thrown. This seems obvious—since the rocket is accelerating, the floor travels upward toward the ball as the ball flies across the ship.

The inside observer sees something different. From his perspective, the flight pattern of the ball is more of a curve; it sort of falls or "drops off" the same way it would if you observed a thrown ball on Earth. In other words, the ball acts as if it were affected by gravity.

Now, imagine we were to do the same experiment but this time with a ball we could throw at the speed of light.

The results we would see would be exactly the same. The observer outside the rocket would see the ball travel in a straight line, whereas the observer inside the rocket would see it travel along a curved path.

Einstein, meanwhile, believed that light was a photon, or a particle. Therefore, doesn't it make sense that in the rocket ship above, light would behave just as the ball does?

genius leaves your head spinning. Instead, turn to Hermann Bondi's *Relativity and Common Sense, a New Approach to Einstein* and David Darling's *Could You Ever?: Build a Time Machine,* which primarily focuses on time travel.

"GO STRAIGHT, YOUNG MAN"

But if you're in a hurry, follow the curved route.

THE ONE BOOK I WOULD READ . . .

I can think of a lot of books that I'm glad I read—not all of them in physics, but books that had an influence on me. I can remember being really shocked as a young fellow by picking up Bertrand Russell and really seeing where it was, and the books by Sir James Jeans, which were semipopular when I was in university. They were not really science books; they were books about science. But they were powerful.

The textbook I always remember is Ernst Mach's *Science of Mechanics.* That's a pretty spectacular book. I suppose if I'd been able to read Latin, some of the older ones would be important, too.

—Richard E. Taylor, winner of the Nobel Prize in physics in 1990

[*The Science of Mechanics* by Ernst Mach] exercised a profound influence upon me . . . while I was a student. I see Mach's greatness in his incorruptible skepticism and independence; in my younger years, however, Mach's epistemological position also influenced me very greatly, a position which today appears to me to be essentially untenable. For he did not place in the correct light the essentially constructive and speculative nature of thought and more especially of scientific thought; in consequence of which he condemned theory on precisely those points where its constructive-speculative character unconcealably comes to light, as for example in the kinetic atomic theory.

—Albert Einstein

Everyone knows that the shortest distance between two points is a straight line . . . but the truth is, it ain't so. And physicist George Gamow did such a wonderful job of explaining why in his book *Mr. Tompkins in Paperback* that we will let him explain:

I suppose that all of you know that a straight line is most generally defined as the shortest distance between two points; it can be obtained either by stretching a string between two points or by an equivalent but elaborate process, of finding by trial a line between two points along which the minimum number of measuring-sticks of given length can be placed.

In order to show that the results of such a method of finding a straight line will depend on physical conditions, let us imagine a large round platform uniformly rotating around its axis, and an experimenter trying to find the shortest distance between two points on the periphery of this platform. He has a box with a large number or sticks, 5 inches each, and tries to line them up between two points so as to use the minimum number of them. If the platform were not rotating, he would place them along a line. . . . But due to the rotation of the platform, his measuring-sticks will suffer a relativistic contraction . . . and those of them which are closer to the periphery of the platform (and therefore possess larger linear velocities) will be contracted more than those located nearer to the centre. It is thus clear that, in order to get most distance covered by each stick, one should place them as close to the centre as possible. But, since both ends of the line are fixed on the periphery, it is also disadvantageous to move the sticks from the middle of the line too close to the centre.

Thus the result will be reached by a compromise between two conditions, the shortest distance being finally represented by a curve slightly convex toward the centre.

WHEW! THAT'S A RELIEF—EVEN EINSTEIN COULD FAIL!

Einstein spent the last 25 years of his life working on what he called the unified field theory.

You can't fault him for trying. After all, you'd think the same guy who was able to show that time and space were one, light and energy and mass were one, and gravity and inertia and acceleration were one would be able to whip together a single theory that would tie the four fundamental forces—gravity, electromagnetism, the strong nuclear force (which is the force that holds subatomic particles together) and the weak nuclear force (which is a form of radioactivity that causes the sun and other stars to shine)—together into one simple theory. But he couldn't do it. And no else has been able to do it, either.

What if there were only one answer to any question you could ask about the physical universe? Well, a unified theory would sure make studying for a physics exam a lot easier, but the reality is that even with all the circular and repeating themes within physics, no one has been able to find one single force that could be the source of it all. There are several almost-but-not-quite booby prizes for unified theorists, however: the electroweak theory, which successfully combines electromagnetism with the weak nuclear force, and the grand unified theory, which attempts to link the strong nuclear force with the electroweak force. So far, however, no one has been able to bring all four fundamental forces (electromagnetic force, strong nuclear force, weak nuclear force, and gravitational force) together into one theory that works.

CONTRAST WITH A THEORY CALLED CHAOS

Physicists talk about chaos theory when they want to explain that any tiny change in one external condition can cause an equally small (or large) change in outcome. For instance, if Tiger Woods makes a miracle putt on the fourth green at the Pebble Beach National ProAm, chaos theory tells us that if one of the spectators had coughed just as Tiger made his putt, the cough might have changed wind conditions enough to alter the course of the ball and cause Tiger to miss the putt. By the same token, if Tiger missed an easy putt on the sixth green, you could also use chaos theory to explain that it wasn't Tiger's putting that was at fault but the heavy breathing of the guy in the front row. Chaos theory was initially introduced to explain the unpredictability of weather—for instance, it was said that global weather conditions are governed by factors that are so sensitive that a butterfly flapping its wings in one part of the world could cause a tornado to occur (or not) in another part of the world.

And you say you want to invite even *more* chaos into your life? Then we suggest you submerse yourself in James Gleick's excellent treatise on the subject—*Chaos: Making a New Science.*

BACK TO THE QUANTUM HIGHWAY

You may recall that Einstein and relativity were just a little side trip off the beaten path of physics. Now it's time to get back to quantum theory, the other story of physics during the twentieth century.

Imagine you are living just before the time of Newton, when there are so many wonderful scientific principles and laws and discoveries to be made—yet those discoveries were there to be uncovered hundreds—if not thousands—of years before Newton found them.

In science, there are times when the work to be done is merely adding to the knowledge that has already been gained. For instance, once gravity was discovered, physicists were kept busy for the next few hundred years just adding more details to what was already known.

Then there are times when discoveries are made that literally change the way people perceive the universe around them. Galileo's findings about inertia constituted one such discovery. Newton's findings about gravity and Einstein's about relativity were others.

Today is one of those times.

Quantum theory is an attempt to explain the behavior of nature's smallest bits, or particles. These particles, hierarchically speaking, are the stuff that atoms are made of—electrons and photons and neutrons. You may have learned in school that atoms are the building blocks of nature. Today, we know that is not true because we have learned that we can break atoms down into something even smaller—tiny bits of matter that we call elementary particles.

As physicists, we know a lot of stuff, a lot of details, about these particles. And we'll share some of that knowledge with you. But in spite of all we know, we have no idea how to take all this information and put it together in a way that allows us to get the big picture. Meanwhile, that statement alone doesn't do justice to our lack of understanding.

If you really want to understand how little we know about quantum theory, think about the Alfred Hitchcock movie *Rear Window.* Jimmy Stewart, you'll recall, plays the role of Jeff Jeffries, a professional photographer who has his leg in a gigantic cast and is confined to a wheelchair. To help pass the time, he uses his telephoto lenses to spy on his neighbors. Early on in the movie, Stewart notices some strange events in the apartment of one neighbor (Lars Thorwald, played by a sinister Raymond Burr).

The neighbor has an invalid wife. Stewart sees the neighbor argue with her. And the neighbor then leaves and returns twice in the middle of the night with a large salesperson's sample case. The next morning, the neighbor's invalid wife is nowhere to be seen, but Stewart can see the man wrapping up a large saw blade and knife in newspaper. The next thing you know, there's a huge crate in the man's apartment and a truck has come to take it away. Meanwhile, the wife is still nowhere to be seen.

Is she dead? In the crate? In the truck? Being hauled far away?

When Jeff cries murder, he is successful at convincing his girl-friend, Lisa Fremont (played by Grace Kelly), of the veracity of the plot, but his friend Tom Doyle (played by Wendell Corey), a detective, dismisses him:

TOM: Lars Thorwald is no more a murderer than I am.

JEFF: You mean to say you can explain everything that's gone on over there and is still going on?

TOM: No, and neither can you. That's a secret, private world you're looking into out there. People do a lot of things in private that they couldn't possibly explain in public.

The similarities between *Rear Window* and quantum theory are uncanny. Here we are with our telephoto lenses (a.k.a. particle accelerators), watching, watching, watching these little particles. But there's nothing we can do without more evidence, without the right evidence. Like Jeff Jeffries, there's nothing more we can do but keep watching, guessing, keep testing. Just as Tom Doyle the detective said, we are peering into "a secret, private world," and particles, like accused murderer Lars Thorwald, do a lot of things in private that we can't possibly explain in public.

A TECHNICAL PHYSICS TERM: "WEIRD"

Who says scientists talk in a way that makes things too complicated? Here is how a few of our fellow science authors have summed up quantum mechanics and quantum theory:

> Quantum mechanics . . . is right. It is also so conceptually weird that physicists to this day feel uncomfortable with it.
>
> —Tony Rothman, Ph.D., author of *Instant Physics*

Quantum theory is correct, and it is as weird as ever.
—Fred Alan Wolf, Ph.D., author of *Taking the Quantum Leap:*
The New Physics for Nonscientists

Relativity is popularly regarded as bizarre and abstruse, but quantum theory is, in many ways, far more so.
—Professor Robert H. March, author of *Physics for Poets*

In his book *Taming the Atom: The Emergence of the Visible Microworld,* Hans Christian von Baeyer unravels the mysteries of the atomic world and shows how scientists are experimenting and playing games with nature's tiniest particles.

TEN RANDOM CLUES ABOUT QUANTUM PHYSICS THAT WON'T HELP YOU UNDERSTAND IT AT ALL

The Photoelectric Effect

If you shine a light at the right kind of metal, electrons will be ejected from the metal's surface. Before 1905, physicists explained this using the principles of classical physics (mechanics + electricity + magnetism): Waves of light would build up in the surface and finally cause an electron to break away. There were some aspects of the photoelectric effect, however, that, given this explanation, still didn't make sense. For instance:

- The photoelectric effect won't occur if you shine the wrong color light at the metal plate. Violet or ultraviolet light works well. Red doesn't work at all. But given an understanding of the photoelectric effect on the basis of classical physics, the color of the light shouldn't make a difference. After all, the only difference between one color and another is frequency. Therefore, the only difference color should make is in how long it takes for electrons to be ejected.
- The rate at which electrons were ejected was determined by the brightness of the light. This didn't make sense, either. When we talk about brightness, what we really mean is that a bright light puts out more energy than does a dim light. And

more energy means a larger amplitude. But why would the height and the frequency of the waves determine whether an electron would be ejected and the rate at which the ejection occurred?

- Meanwhile, even though brightness increased the rate of the ejections, it didn't have any impact on the energy of the electrons once they were free. Yet in classical mechanics, the way to increase energy is to increase brightness.

- Finally, it was discovered that if you wanted to increase the energy of the electrons ejected in the photoelectric effect, you had to change the frequency of the light. Back to red versus violet versus ultraviolet—which doesn't change the number of electrons you get, but it does change their energy. Again, this is contrary to what classical mechanics would predict.

Einstein, in his miracle year, finally came up with the solution. He did it by building on what he had learned from Max Planck.

You'll recall that Planck said that energy in an electron worked sort of like a vending machine. If you have enough change, you get energy. If you don't have enough change, you get nothin'.

Einstein took Planck's theory a little further by asking what happens if light is a particle. If it is a particle, it is moving. If it is moving, it has energy. If it has enough energy, we get the photoelectric effect. If it doesn't have enough energy (for example, red light) you get nothing. If you increase the frequency, you don't get more electrons—you either get one or none. However, increasing frequency will increase the energy of the electron. On the other hand, if you increase the brightness, you don't get more energy but you do get more electrons because if light is a particle instead of a wave, then increased brightness means more particles . . . which means more electrons can be ejected.

Suddenly, the life span of the wave theory of light was looking very short.

Quantum Leaps

More weirdness. As we mentioned earlier, Ernest Rutherford had figured out that the atom is mostly space with something tiny (which he called the nucleus) at the center.

Well, obviously all the atoms of the world are not burning up and collapsing, so there must be some other thing going on that wasn't yet understood . . . that is, until Niels Bohr of Denmark went to work on the problem.

He accepted Einstein and Planck's notion about quanta but only partially accepted Rutherford's picture of the atom. Instead of a solar system approach in which electrons are seen as always orbiting around something at the center and always emitting energy as a result (much like we think of solar systems in classical mechanics), Bohr asked what would happen if we had a "solar system" in which electrons emitted energy only when they moved from orbit to orbit.

It was a crazy idea. There was absolutely no precedent for it. But he was right.

When the electron jumps from one orbit to another, this is called a quantum leap. Don't ask how the electron does it or why it does it or even how it can do it. No one knows the answer to those questions.

But here's what we do know: The energy level of an electron depends on which orbit it is on; the closer the orbit is to the nucleus, the lower its energy. Therefore, when electrons jump, they will leap to an orbit level that is nearer to the nucleus. Otherwise, if you throw them a little energy to stimulate them, they'll go to a higher energy orbit farther from the nucleus. Regardless of which way they go, there is always a change in energy. And that energy has to go somewhere . . . so the electron emits it in the form of a photon. And the word *photon*, you'll recall, is really just a word Einstein uses to describe individual particles of light.

Understanding Bohr's picture of the atom makes it easier to understand Planck's notion that an electron will emit energy only when it changes energy in a jump, since an electron's movement from orbit to orbit occurs in finite steps (like a staircase), not on a continuum like a ramp. The electron must have enough energy to make it up the next step. If it does have enough energy, it will then emit what is left over, since it will always go from a higher energy level to a lower one with some energy left over.

Meanwhile, we can also use what we just learned about the quantum leap to better understand the photoelectric effect. When we shine the light on the metal plate, what we are really doing is shooting a stream of photons—light particles—at it. The metal plate is covered with electrons. They are held in place by electrical forces. To break away, they need enough energy to make a quantum leap away from the plate.

As we now know, this requires a specific amount of energy—again, the correct change. When the light shines onto the plate, the photons either hit electrons or they don't.

To illustrate, let's take a single photon—call him Jack. If Jack is lucky, he'll fly out of the flashlight and strike an electron. Let's call the electron Jill.

If Jack is a photon of red light, the story ends. He isn't of a high enough frequency to fall down the hill and get Jill adequately excited. But if Jack is of the ultraviolet frequency, for instance, he transfers some or all of his energy to Jill, and this energy will be sufficient enough to cause her to fly off the plate.

Identity Crisis
. . . or . . .

A wave is a wave is a wave is a wave is a wave is a wave is a wave . . . is a particle.

Put Einstein in the ring against Newton. Isaac says light is a wave, Al says light is a particle. On second thought, let's be pacifists and call for a compromise; let's say light is a wave *or* a particle. Or both at the same time.

Obviously, the photoelectric effect demonstrates that light is a particle—because the transfer of energy from the photon to the electron is in discrete amounts and occurs in much too concentrated an area to be a broad, expansive wave. On the other hand, as we mentioned earlier, a guy named Thomas Young—way back in 1800—came up with a way to prove that light is a wave.

Here's what he did: He directed a light toward a screen with a slit in it. Behind the screen was another screen with two slits in it. Behind the screen with two slits was a third screen with no slits. If light is a wave, it should squeeze through the slit in the first screen and then fan out again in a new set of waves . . . only to reach the second screen. There, it should again squeeze through the two slits and then fan out into two sets of waves and create an interference pattern on the final screen.

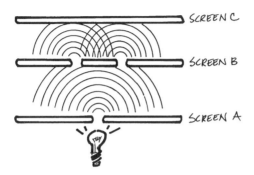

Of course, the result of Young's experiments were exactly as predicted, and he proved that light is a wave. (The proof really lies in the fact that particles don't cause interference—therefore, the interference pattern on the last screen had to be the result of waves of light.)

More than 100 years later, Einstein wanted to prove that light waves can act like particles and, of course, someone else (a guy named Louis Victor de Broglie) caught on and tried to prove that a particle could have a wavelength.

Both, of course, were right, or we wouldn't have brought it up.

What is interesting is that the light wave–versus-particle identity crisis was clarified by again using Young's double-slit experiment . . . with a few slight variations:

First, let's cover the last screen (the one without any slits) with photographic paper. Then, let's eliminate the first screen (the one with only one slit). Then let's cover one slit on the second screen. Now, when we shoot our photons at the screen, those that pass thorough the slit make a predictable pattern of particle hits on the photographic paper when they hit. Next, because we want to be thorough, let's open the slit we covered and cover the slit we left open, and then repeat the experiment. As expected, we get the same, predictable pattern.

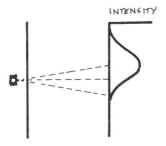

Electrons reach the photographic paper through one slit in a screen.

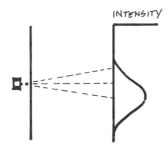

The same pattern of electrons is displayed on the photographic paper when the electrons are allowed to enter a different slit in the same screen used the first time.

Now the fun begins!

What happens if we shoot photons at the screen with both slits open? Well, what we would expect is a pattern of hits on the photographic paper that is the total of the previous two trials.

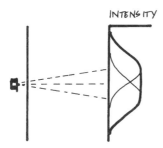

This is the pattern of electrons we *expect* to be displayed on the photographic paper when there are two slits open in the screen.

Given these logical expectations, however, we would be stunned by the actual results. Instead of the predicted outcome, we get a pattern of hits on the photographic paper from our particles that looks almost exactly the same as the interference pattern Young got when he did the experiment with waves of light.

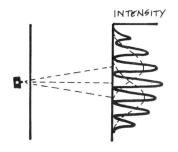

This is the pattern of electrons that is *really* displayed on the photographic paper when there are two slits open in the screen.

How can we explain this?

We can't. Here is where quantum physics goes from being mildly mysterious to completely, outrageously weird. We can talk about it, argue about it, and wish the results were different, but they're not, and for now, we don't know why. In fact, right now the best we can do is interpret what we think the results mean and start ruling out some interpretations:

Q: Okay, particles don't experience interference, but they can bump into each other, right?

A: Yes, they can bump into each other, but that's not what's happening here. To rule this explanation out, scientists have repeated the experiment umpteen times in such a way that only one photon at a time was shot at the screen and no more photons were released until that photon hit the photographic paper. In every case, the results were the same.

Q: Is it possible that the photon splits itself into two and goes through both slits?

A: Nope. The experiment has been repeated umpteen more times in such a way that only one photon is shot at the screen at a time and as each photon goes through a slit, an alarm is sounded. There was never a case where something went through both slits after a single photon was released.

Q: Now, let's up the weirdness ante. What if we set up a sensor between the two screens so we can see the photons as they pass through one slit or the other?

A: In theory, this should work. Just as each photon comes through one slit or the other, it will be sensed and we will be able to see which slit it came through . . . or we should be able to see how a single photon somehow manages to come through both slits at the same time. But it *doesn't* work. Every time a photon passes through a slit, we can see where it is and record which slit it has passed through. This is great news, except . . . the pattern that this set of photons leaves on the photographic paper is different than the pattern we got in the earlier experiments. No longer does it appear that the photons are actually waves passing through both slits. Now, the pattern is exactly what we predicted it would be before we had a chance to run the experiment. The result is the total of the two patterns we got when we shot the photons at the screen with only one or the other slit open. And this, of course, raises an entirely new question:

Q: Could it be that the sensor itself affects the photons as they fly through it?

A: This is a good question, and it has led to umpteen more sets of experiments. The conclusion is that there is no way we can arrange the sensor so as to watch the photons without having some kind of impact on the final results. No matter what we do, this experiment brings us this disturbing set of results: If we try to watch what the photons do, we get a different result than if we don't try to watch.

Are you, too, wondering why particles do the things they do? In his book *Taking the Quantum Leap: The New Physics for Nonscientists*, Fred Alan Wolf does an excellent job of explaining what he calls the new physics.

Werner Heisenberg's Uncertainty Principle

Obviously, we can't leave things as they stand—with particles that act like particles when you send them toward a screen with one slit but act like waves if you send them toward a screen with two slits. How do they know what to do? And what difference does it make to them whether there is one slit or two? And how do they know we're trying to catch them when the apparatus we use doesn't do its job until after the electrons have gone through one slit or the other? Do they somehow communicate with the others back at home base? Or do they have ESP (extrasensory perception)?

Scientists didn't know back in 1927, and we still don't know now. But they did know something, which Werner Heisenberg was able to boil it down into one law which he called the uncertainty principle.

Perhaps the choice of names was unfortunate. To many laypeople, it is something to mock. Contrary to what its name implies, the uncertainty principle is quite explicit about what can be known and what can't be. Here it is from the horse's—or Heisenberg's—mouth: "It is impossible to simultaneously measure the position and the momentum of atomic particles with an arbitrary degree of accuracy"—which may cause you to experience some uncertainty as to what it was that Heisenberg meant.

So let's make it simple. You are the left fielder for the New York Yankees. Ken Griffey Jr. of the Seattle Mariners is at bat and swats one hard. The ball arches up high into the air and is headed in your direction. You don't think; you move. And you plant yourself on the ground right where the ball is going to land.

Well, what if that baseball had been an electron? What are the odds that you would have caught it?

Pretty slim.

As we learned in the double-slit experiment, if we want to see something, we have to expose it to light. And when we expose something to light, what we are really doing is bombarding it with photons. And those photons take their toll. Of course, the impact of the light glancing off a baseball is minuscule. But the impact of those same photons bouncing off an electron is great.

Although Heisenberg didn't explain his uncertainty principle using a baseball analogy, what the principle expresses is the same. When the baseball comes toward you, you know how to estimate its speed and its path so as to place yourself right where the ball will go.

What Heisenberg is trying to tell you is that when you're out in left field and trying to catch an electron, you can either know its speed or you can know its position . . . but you can't know both, for one of them will automatically be altered by virtue of the fact that you are looking. After all, to look at the ball (or an electron) causes it to be bombarded by photons. Although those photons may have no effect on a big, fat baseball, they will have a dramatic effect on a tiny little electron—and that effect will either alter the speed or the path of the electron as it flies through the air.

In either case, you've got a problem, because you won't know how far forward or back to run or how far to the left or right to run, because you don't have enough information to make an accurate judgment.

Perhaps, on this level, soccer provides a better analogy. You are downfield near your opponent's goal, your teammate kicks the ball to you, you set yourself up in a position to receive the pass so you can kick it into the goal when—suddenly—an opponent leaps high into the air and butts the ball with his head. In this analogy, the ball is like an electron and your opponent's head is like the bombarding photon. And once he makes contact with the ball, it is anyone's guess as to where it will go—and you are no longer able to predict both the ball's position and its momentum.

Heisenberg drives his point home: On the atomic level, we cannot predict the future because we have no way of gathering all the facts. Instead, all we can do is discuss the odds or probability that a particular event will occur.

Observers Versus Participants

Let's go back to our baseball example. Again, you are the left fielder. Again, Ken Griffey Jr. whacks one out over your head . . . at first, you are an observer of the ball. You watch it and gauge its speed and position. Armed with that information, you know exactly where to run to catch it.

Now, imagine again that it is an electron you want to catch.

You watch the electron as it sails over your head, and you gauge its speed and position. Armed with that information, you know exactly where to run to catch it . . . except that at the moment you look at the ball, by virtue of the fact that you are looking, the electron is bombarded with photons and its speed or position changes. You have no idea where to run, so you miss the catch.

In the first example, you are an observer. You watch the ball. You run and catch it.

In the second example, you are a participant. You watch the electron. By virtue of the fact that you are you watching, the behavior of the electron changes. You suffer the consequences of Heisenberg's uncertainty principle.

You also learn, firsthand, about a major difference between classical and quantum physics. In classical physics, we learn about the physical world by observing it—and, for the most part, that which we are observing is undisturbed by our watching eyes.

It's a different story in the quantum world. We are not observers. We are participants. By virtue of simply looking, we alter the future. But if we want to understand, we have to look. But by looking, we don't really get to see—or at least we don't get to see what is there when we don't look, and therefore we can't make perfectly accurate predictions, anyway.

Erwin Schrödinger's Cat

No one has ever summed up the weirdness of quantum physics better than Erwin Schrödinger did with his paradox about a cat in a box. As you are surely beginning to see, the nature of activity on the atomic level is unlike anything you have ever seen or imagined.

BUT WHAT ABOUT SCHRÖDINGER'S CAT?

Erwin Schrödinger didn't like Werner Heisenberg's uncertainty principle. One way he demonstrated that was by coming up with his own theory that—cruelly—involved a cat.

The theory works like this: We take a perfectly good cat and put it in a cage with a single atom and a diabolical contraption that will release deadly gas should the atom strike it. If you leave the cat in the cage for 1 hour, there is a 50-50 chance the atom will strike the vial of gas.

When you come back in 1 hour, your mind contains one image of a happy, living cat and another image of a dead, stiff cat. Yet what you know before opening the cage has no relationship to the actual outcome inside the cage—and this, for Schrödinger, was a problem because if you followed the logic of Heisenberg, the cat would be neither dead nor alive until the observer opened the cage and looked.

AND EINSTEIN'S CAT . . .

The wireless telegraph is not difficult to understand. The ordinary telegraph is like a very long cat. You pull the tail in New York, and [the cat] meows in Los Angeles. The wireless is the same, only without the cat.

—Albert Einstein, on the latest rage in technology of his day

In the early years of quantum physics, while all the players were throwing out wild and improbable ideas in an attempt to explain what was going on, Schrödinger tried to bring things back to more stable, predictable ground by suggesting a mathematical scenario that removed much of the oddness of the wave–particle duality and brought the activity of the particle back to a seemingly Newtonian basis.

How'd he do it? Well, the math is much too daunting to explain here, but in the most simplistic terms he began with de Broglie's theory that every particle acts like a wave and he started building from there.

Schrödinger's theory was this: If a particle acts like a wave and waves cause interference, then there ought to be a way to mathematically chart the whereabouts of the wave. And if you can chart the whereabouts of the wave, then instead of looking at that wave as a moving chunk of matter or a field, you could also look at it as a probability wave. It will give you a distribution (sort of like the pattern you get when you fire a shotgun shell at a paper target) that indicates the probability of where the electron is likely to be at any given time.

Now, this sounds pretty wild when compared to the world as Newton described it, but on the other hand, when you compare Schrödinger's picture of reality on the subatomic level to the pictures painted by Heisenberg and the double-slit experiments, Schrödinger's is pretty tame. After all, waves were a known quantity. Particles were a known quantity. And to the old guard (like Einstein, Planck, and even Schrödinger himself), the fact that a more tangible explanation could be found to explain the behavior of matter on the subatomic level was a giant relief.

History Repeats Itself (Again)

What a relief that we no longer persecute physicists when we disagree with them—as happened with Aristarchus (who was persecuted by the Stoics), Bruno (who was burned at the stake), and Galileo (who spent the last 10 years of his life under house arrest). Today, we're a little more democratic. We talk things over and then, by consensus, we crown as the short-term winner the one who gets the most positive press. The long-term winner, on the other hand, is always the one who can come up with the most compelling evidence or proof.

In the history of physics, the two biggest winners have been Isaac Newton and Albert Einstein, of course. Both had great press and very little bad spin.

But very few people remember how upset Einstein was about the direction in which quantum physics was headed. He didn't like the lack of determinism, the dependency on faith and probability and guessing that Bohr and Heisenberg and others were advocating. In fact, it became a heated battle, with he and Planck and Schrödinger in one corner and the aforementioned Bohr and Heisenberg in the other. Einstein saw Schrödinger's wave function as a happy medium. Waves, he understood. The probability stuff, he could live with. Except the other camp wouldn't let him. They hammered away at the old view of Einstein and his camp, who continued to come up empty-handed. They couldn't supply any evidence of the old tangible mechanical world as it applied to the subatomic zone because, well, there wasn't any such compellingly conclusive evidence to be had.

YOUR FAVORITE PHYSICISTS TRADE BARBS

Albert Einstein to Niels Bohr, in an effort to refute the uncertainty principle: "God does not play dice!"

Bohr's retort: "Einstein, stop telling God what he can and cannot do!"*

How a beaten Erwin Schrödinger finally gave up the fight against uncertainty: "If one has to stick to this damned quantum jumping, then I regret ever having gotten involved!" (Ironically, it was Max Planck's vending-machine approach to quanta and Schrödinger's wave mechanics that finally made it possible for Bohr and the other advocates of quantum mechanics to prove their theory. Meanwhile, Planck and Schrödinger regretted their inadvertent contributions to the unpredictable quantum physics for the rest of their lives.)

---

* One must wonder if Bohr was making sly reference to one of Einstein's comments seven years earlier at news that British astronomer Arthur Eddington had confirmed Einstein's theory of general relativity. When news of the confirmation reached Einstein, he was asked what he would have done if Eddington proved his theory of relativity to be wrong. Einstein replied, ". . . I would have felt sorry for the dear Lord, for the theory is correct!"

"How Much Is That Particle in the Window?"

Just as your body is chock full of organs you've heard of before but rarely (if ever) give a second thought to, the universe is similarly full of subatomic particles. Surely you've heard of many of them, yet it's doubtful you could explain the difference between them. Until now:

- **Electron:** The first particle to be discovered. Has a negative charge. These are the guys that make quantum leaps from one energy level of the atom to another. In other words, it is because of the electron that all this quantum trouble began.
- **Proton:** The electron's stay-at-home counterpart. They live in the nucleus of the atom and come with a positive charge, perhaps from listening to too many Tony Robbins motivational audiotapes.
- **Neutron:** Remember the "plum pudding" model versus Ernest Rutherford's "solar system" model? It was the neutron, a particle with no charge, which was at the center of the debate. The neutron and the proton account for most of the weight of an atom—in fact, the atomic weight of an atom is the sum of the total number of protons and neutrons. By way of comparison, both the proton and the neutron are about 1,800 times heavier than an electron.

You may recall that in the beginning of this book we said it is always the goal of the physicist to end up with the smallest number of rules to describe the greatest number of phenomenon. In physics, tidy is nice.

For that reason, the picture of the cell with an electron, a proton, and a neutron was a pleasant one. You've got something on the inside, something on the outside, and a little something that flies off and makes light. And you've got a particle with a positive charge, another with a negative charge, and a third that is neutral. From the perspective of simplicity and natural beauty, what more could you want from life?

Yet once again, the universe defies humankind's desire for tidiness. To date, there is no set of simple laws that provide a solution to as many questions as Newton's laws of motion and gravity did for classical physics. Instead, we get a universe full of so many particles it's hard to remember which is which.

- **Positron:** Also known as the antielectron because it's the spitting image of the electron, except it's got a positive charge.

- **Neutrino:** You go on vacation for a month. Conveniently, you leave the day after the meter reader comes to see how much electricity you used the previous month. Equally convenient, you and the reader both return to your house on the same day a month later. Before you left, you made sure you turned all the electrical appliances off, yet while you were gone, you still somehow managed to burn $24 worth of fuel. In the 1920s, physicists had a similar situation. They knew that certain radioactive elements decayed by emitting electrons. They knew how much energy the cells were emitting and how much energy was being carried away by the electrons. And there was a whopping difference between what the cells were emitting and what the electrons were carrying away. So where was the additional energy going? Off with another particle, called the neutrino.

- **Muon:** With a mass 200 times greater than that of an electron, but who's counting?

- **Pion:** One hundred and forty times heavier than the electron.

- **Lambda, kaon, hyperon, sigma plus, sigma minus . . . :** And there are at least 100 more. Ultimately, it sort of becomes like filing—if you're going to have 100 different kinds of building blocks, then you've got to figure out some way to classify them so they fit into smaller groups. Of course, if you're going to start classifying things in physics, the first place to start is to break the particles down into a group associated with matter and a different group associated with forces.

- **Fermions:** This is the stuff of which matter is made. These are also the particles that have antiparticles associated with them. (Antiparticles have the same properties but a different charge. These pairings include the following: electron and positron, proton and antiproton, neutrino and antineutrino, and the quark and antiquark.) Antiparticles are usually found in cosmic ray showers or produced in a particle accelerator.

Fermions can be further divided into two groups: (1) hadrons, which are composed of quarks, the quark being an elementary particle—in other words, it can't be broken down any further . . . yet; and (2) leptons, which include the electron, the muon, the tau, and three different types of neutrino—all leptons are elementary particles, as well.

■ **Bosons:** These are known as mesons or the particles associated with different forces, and all are made from the combination of a quark and an antiquark.

■ **Quarks:** These are kind of fun, in and of themselves. They come in six varieties: up, down, top, bottom, strange, and charm. They also come in colors—red, blue, and green. And, of course, each of the above has its own antiquark, making for the antiup, antidown, antitop, antibottom, ad nauseum.

How much more about particles do you really want to know? How about subcategories of subcategories? Baryons are heavy particles like neutrons and protons. All baryons are fermions . . . but not all fermions are baryons. Mesons are middleweight particles like the pion. All mesons are bosons, but not all bosons are mesons. Leptons are light fermions, like the electron, the muon, and the neutrino. Finally, hadrons are all baryons and mesons.

■ **Graviton:** Of course, the ultimate particle associated with a force is the one that is associated with the law of gravity—the graviton . . . but no one's been able to prove its existence yet.

Finally . . . Feynman

So, we've got all these particles. And still, the most basic question is this: How does this quantum jumping thing work? How does it happen in a way that is both consistent with the squirrellyness of Einstein's relativity and the weirdness of quantum mechanics? The answer to that question is a theory called quantum electrodynamics (QED).

But finding the answer was a long, drawn-out affair that brought progress in quantum physics to a virtual standstill for longer than a decade. The physics community knew these particles existed and continued to discover even more of them, but they couldn't explain the most fundamental interaction between them.

It was Richard Feynman's father who asked the question in such a way so as to make the problem crystal clear, as Feynman recounted in his book *What Do YOU Care What Other People Think?*:

Once . . . when I came back from MIT (I'd been there a few years), he said to me, "Now that you've become educated about these things, there's one question I've always had that I've never understood very well."

I asked him what it was.

He said, "I understand that when an atom makes a transition from one state to another, it emits a particle of light called a photon."

"That's right," I said.

He says, "Is the photon in the atom ahead of time?"

"No, there's no photon beforehand."

"Well," he says, "where does it come from, then?"

Although Feynman died at the age of 70 in 1988, he was portrayed in a one-man show in New York City in 1994. Early in the show, the actor playing Feynman posed the same question to the audience and explained why physicists had had such a terrible time trying to find the answer:

I, Feynman, am standing here. You, madam, are sitting there. The seat you are sitting on is beneath you. We can point to it. We can touch it. We know it is there, because it is there. In fact, we can tear off a piece of wood from your seat, and even if it were so small that you needed a microscope to see it, we'd still say, "Here it is!"

If we wanted, we could give that tiny piece of wood a location, an exact position. We can say: "This microscopic piece of wood I'm holding in my hand has the coordinates in space of 40.438 degrees latitude and 74.195 degrees longitude." And if you got out a map and figured out where those precise numbers and coordinates were on the planet, you'd find that held up in midair by me, Feynman, was this teeny-weeny bit of wood right here.

Okay, so that's easy.

Now, let's take that same piece of wood and look at something smaller in it, its atoms; then, let's look at the smallest things that it's made of—its particles. Let's point to those pieces, the particles, and give them a precise location or coordinates, too.

Well, this is where it gets really weird: We can't do it!

The better our machines get at detecting and observing particles of nature, the worse we got at saying really where those particles are.

Where they are in space, or if they are there at all. There's the wood—a tiny splinter—it's in my hand. It's there; we can see it. Right here! The wood is made of particles. So why we can't say that the particles that make up the wood splinter are actually here?

All our calculations, all our formulas kept telling us that the position of those particles were in any number of places—infinite places—all at the same time! What the hell was going on?

It would be as if the chair that you're sitting on was not only holding you up right there, but that the chair was also over there and over there and over there . . . all at the same time! We could no longer say with certainty that that chair was there. Or rather, that the electrons and other particles that make up the chair are there . . . or anywhere. You see, the smaller and smaller and smaller we got, looking for the most fundamental stuff that nature is made of, the slippier everything got . . ."

Near the end of the show, Feynman explained how he found the answer to this riddle—for which he won the Nobel Prize—and revealed to the audience a simple truth about physics:

A few years after World War Two, I began to wonder why I couldn't figure out the secret of these particles by actually imagining that I was one of them . . . so I started trying to think like a particle. "Think like a particle, think like a particle, think like a particle," I'd say to myself. "If I were a particle, what would I be like? What would I want to do with my life?"

Ok! So I'm inside my head and in my head I'm seeing . . . I'm seeing what it's like in there, inside the atom.

We've got matter. Matter is made of atoms. Around those atoms are electrons moving around in all these crazy directions. Now, we knew that electrons were fundamental—indivisible— and that matter doesn't break down any further. So we got this tiny piece of matter—the electron. And that electron is moving from place to place . . . yes, I was seeing something in space and time. There were quantities associated with points in space and time, and I would see electrons going along—

—and let's see, I knew an electron emits a photon—Einstein told us that. Anyway, pow!: Photon comes zipping out of an electron

—and it's got these wave-like properties, even though it's a particle, too. And you see, when the photon comes out, that electron goes peeeerrrww! Off in another direction.

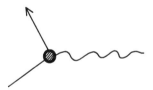

Well, I'd been doing these path integral ways of doing electrodynamics for so long, and thinking about it for so long, visualizing in space and time these things, working out perturbation expressions so many times, and I'm not . . . I'm not doing half advanced and half retarded, but the full retarded and all that stuff.

And you've got to take into account that there are other electrons out there at the same time—

—then, I had this half-assed idea—see, I wanted to put in one of my fundamental gut intuitions: that in the quantum world, all actions of the players—the particles—was really interactions. . . .

That deep down in nature, the most important thing that could be said about the smallest pieces of matter is that they're interacting with each other; they're wiggling; they're jiggling; they're doing this dance with other pieces of matter, other particles, other electrons—and it's all the same stuff—they're all doing the same thing. . . .

So, this electron absorbs the photon. Sucks it right in. Which disturbs the original path and direction of our first electron and causes it to take off again to another route . . .

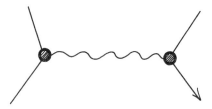

. . . where it starts the process all over again and again and again and again and again and again and again and again and again, all of this happening trillions of times, even in a single quick flash of light!

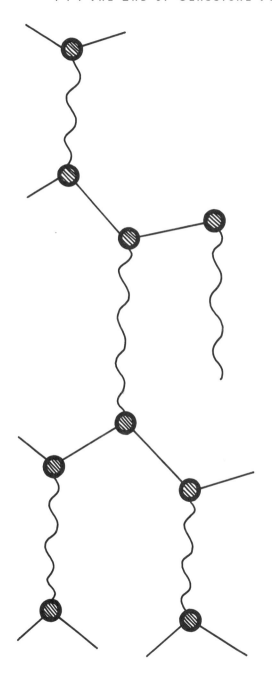

I got it! I drew a picture of the electron inside the atom!

It was for answering the QED question that Feynman (and two other physicists, Julian Schwinger, who had been Feynman's intellectual rival since high school, and Shin'ichirō Tomonaga of Japan) eventually won the Nobel Prize.

What set Feynman apart from the other two physicists was that he took a graphic—instead of a mathematical—approach to the problem. Today, his methodology, using a type of graph that is now known as a Feynman diagram, is a tool that physicists and others in the sciences use to solve many many types of problems, not just those relating to quantum physics.

Just how valuable are Feynman's diagrams to other scientists? We'll let Freeman Dyson, another physicist, explain:

> We [Dyson and another physicist, Cécile De Witt] spent the weekend talking with Feynman, and we put these two problems to him, and asked him, "Look, how do you deal with those?" Feynman simply said, "Oh, we'll see about that." He proceeded, in front of our eyes, to calculate both these problems and get finite and sensible answers for both of them, showing that the thing [his diagrams] really worked, even for the hardest cases. That was just about the most dazzling display of Feynman's powers I've ever seen. These were problems that had taken the greatest physicists months to fail to solve, and he knocked them off in a couple of hours.

And in 1985—50 years after he developed his first Feynman diagrams—Feynman had this to say:

> The theory of quantum electrodynamics has now lasted for more than fifty years, and has been tested more and more accurately over a wider and wiser range of conditions. . . . To give you a feeling for the accuracy of these numbers, it comes out something like this: if you were to measure the distance from Los Angeles to New York to this accuracy, it would be exact to the thickness of a human hair. . . . We physicists are always checking to see if there is something the matter with the theory. That's the

game, because if there is something the matter, it's interesting! But so far, we have found nothing wrong with the theory of quantum electrodynamics. It is, therefore, I would say, the jewel of physics—our proudest possession.

It's worth noting, by the way, that in Latin, QED means *quad eret demonstratum,* or "We have shown what was meant to be shown." In other words, here lies everything there is to be known about the subject at hand. For that reason, mathematicians and scientists often finish their research off with the acronym *QED.* And given how long Feynman, Schwinger, and Tomonaga's theory about light has been with us in spite of all attempts to prove it wrong, both meanings of *QED* may apply.

That Was Then, This Is Now

It's interesting to note that although QED may be the crown jewel of physics, there are still a lifetime of questions to be answered. Feynman, Schwinger, and Tomonaga made it possible to understand and predict the behavior of nature's smallest particles when they participate in the most fundamental interaction . . . but what about more complex interactions, like the behavior of the electron in the double-slit experiment? How much progress has been made there? Lots, yet also very little.

. . . nobody understands quantum mechanics.

—Richard Feynman

And no one explains quantum mechanics better than Feynman did.

One story we find amazing is about two physicists, David Wineland and Chris Monroe, at the National Institute of Standards and Technology in Boulder, Colorado. These two scientists have successfully managed to get a single atom to exist in two different places at one time, as was reported in *Discover* magazine in 1996:

David Wineland and Chris Monroe pulled off this feat using lasers and a magnet to manipulate a beryllium atom inside a vacuum chamber. They first confined the atom inside an electromagnetic field and, with lasers, bounced photons off it until it rested essentially motionless. Using another laser burst, they pumped just enough energy into the atom so that it had an equal chance of assuming either of two quantum states known as spin-up and spin-down, which describe the orientation of the magnetic field of the atoms electrons. Just as with [Schrödinger's] hapless cat, the atom, until it is actually measured, exists simultaneously in both states.

Physicists have been creating such odd, mixed states within atoms for years. But what Wineland and Monroe did next was unprecedented. They calculated that a light pulse with a wavelength of exactly 313 billionths of a meter, and of a precise polarization (which describes the direction in which a light wave vibrates), could move the atom in its spin-up state without affecting the spin-down version of the atom: atoms in different quantum states absorb only very specific wavelengths and polarizations of light. The right light, in other words, enabled Wineland and Monroe to tease apart the superimposed versions of the atom. . . . Wineland and Monroe's research, as esoteric as it seems, may one day find a practical application. The two physicists are interested—as are many researchers—in the feasibility of something called a quantum computer. Atoms in such a computer would replace transistors and other electronic components, greatly shrinking the size and increasing the power of computers. In a quantum computer, one atom could simultaneously represent a zero and a one in the binary language of computers.

HOW TO MAKE A STARTLING NEW DISCOVERY IN PHYSICS

If not you, then who?

After all, Einstein was originally an outsider. So was Newton. And someone's got to figure this quantum thing out.

Here's all you've got to do:

HOW TO SUCCEED IN PHYSICS

#1: The Right Way	#2: The Fast Way	#3: How to Generate a Spontaneous Stroke of Brilliance
Tycho Brahe knew the right way to make a startling discovery, and he also knew that it would involve a lot of painstaking observation and meticulous recordkeeping.	The fast way is to let someone else do all the work while you steal all the credit. Or, to be more accurate, let someone else do most of the work and get stumped, at which point you step in with the missing link that solves the entire puzzle.	Let greed inspire you! Isaac Newton developed, wrote, and published his laws of motion in less than three years once Christopher Wren offered a cash prize to the first person who could explain how Kepler's laws worked and why they were correct.
Like many other astronomers and physicists of his time, Brahe was fascinated and perplexed by the movement of the planets. But he wasn't content to sit around and engage in a philosophical debate about it the way others like Aristotle and Ptolemy and Copernicus had been content to do. Instead, he decided, he would figure out what made the planets tick by keeping track of their every movement.	No one better embodies this spirit than Johannes Kepler, Tycho Brahe's assistant.	Let the opposite sex inspire you! Richard Feynman claimed that he did his best thinking while sitting at a table in the back of a local topless bar.
For 20 years, he monitored the positions and movement of the planets and kept flawless records on what he saw—and he made some significant contributions and discoveries. Most notably, he brought a degree of sophistication and accuracy to	As you'll recall, it was Brahe who slaved and toiled to collect data about the heavens above. But before he could use the data to come to any definitive conclusions, he suffered an untimely death, which left Brahe's assistant, Kepler, sitting in the driver's seat. Kepler took over Brahe's job—and the data—and used both to unravel the elliptical nature of planetary motion.	Let bad television inspire you! What better way to get the mental juices rolling than to watch television's worst detective solve the same boring old cases over and over again? It worked for scientist, historian, and science fiction great Isaac Asimov, who once had this to say about Columbo: "My all-time favorite show is Columbo. I love it. I love the acting. I love

#1: The Right Way	#2: The Fast Way	#3: How to Generate a Spontaneous Stroke of Brilliance
astronomy that was unprecedented.

For instance, he made it possible to measure the movement of the planets more accurately by making bigger instruments. In that way, he could divide their scales into smaller and smaller increments.

Furthermore, Brahe demonstrated that it was important to observe a subject before forming theories about it, and he demonstrated that the way in which observations are made is also important. In his time, it was standard for astronomers and physicists to watch each planet at key times (such as when a planet is in opposition, where the Earth lies directly between the planet and the sun), but Brahe insisted on observing the planets during all phases of their orbits. As a result, he observed some patterns in the behavior of the planets that had never been seen before:

■ That the planets orbit the Sun
■ That the shape of their orbits is an ellipse | | the mystery. I love everything about it. I sit and watch it over and over again." |

#1: The Right Way	#2: The Fast Way	#3: How to Generate a Spontaneous Stroke of Brilliance
■ That the Sun is located at one focus of each planet's orbit Ironically, however, it was not Brahe who would finally discover these three laws. Instead, that honor would belong to his own assistant, Johannes Kepler. But you can read more about that in the next column.		

ARE GREAT PHYSICISTS BORN OR MADE?

To find out, we asked five winners of the Nobel Prize in physics what they wanted to be when they were growing up.

- **Arthur Schawlow, 1981 winner of the Nobel Prize in physics:** "My mother says I once said I wanted to be a garbage man. At that time, I think it was before we left New York. I must have been about three or less, but the garbage men there wore white uniforms and they had white trucks and I thought they looked real glamorous."

- **Philip W. Andersen, 1977 winner of the Nobel Prize in physics:** "Most of the time, I think it was a scientist. . . . I certainly never wanted to be a fireman or any of those kinds of things. There was a period in college when I wanted to be a free-verse poet, but you get over that."

- **Jerome I. Friedman, 1990 winner of the Nobel Prize in physics:** "I started out . . . in high school as an art student. I enjoyed painting. I went to a high school in which I was an art student in a special program, so I used to paint for three or four hours a day and took other courses during the other hours. . . . I think I would have been happy as an artist [but] I probably wouldn't have been, maybe, a very successful artist."

- **Burton Richter, 1976 winner of the Nobel Prize in physics:** "I wanted to be a scientist. I knew that—oh, I would say—from about the time I was ten years old. But it wasn't until I had my first physics courses and chemistry courses at MIT that it became very clear to me that physics was what I wanted to do because I got to feeling then that chemistry was kind of like cooking. . . . My chemistry friends would be very angry at that characterization, but that's the way I felt."

- **Richard E. Taylor, 1990 winner of the Nobel Prize in physics:** "I assumed that I would probably like to be a surgeon, but my grades were not nearly good enough to go into medicine. Actually, I blew my fingers off in an accident. Later, this doctor that lived next door said to me, "Look, fella, people aren't going to have much confidence in a surgeon with half his hand gone," and I said, "Gee, I never thought of that."

THE DIFFERENCE BETWEEN SCIENCE AND TECHNOLOGY

. . . and why Bill Gates hasn't been successful at either. There is a vast difference between science and technology. Science is the pursuit of knowledge for the sake of knowledge. Scientists are driven to find answers because, darn it, there are still elements of the universe out there that we just don't understand. Technology, on the other hand, is the pursuit of knowledge for humankind's sake. Technologists are driven to find answers because, darn it, there's money out there to be made. And then there is Bill Gates:

- He made his mark on the computer world with a software called DOS (for Disk Operating System). Bill didn't write DOS. He bought it for a song from a college professor who wasn't smart enough to realize the potential of what he had.

- DOS was such a difficult and cumbersome program—particularly when compared to the cute, fun, and easy-to-use operating system that Apple computers had—that Gates knew he was going to be up a creek unless he did something drastic. Thus, Gates's own easy-to-use Microsoft Windows was born—and it allowed Gates to cover up the difficult DOS interface as if he were redecorating a room.

- From a sophisticated user's point of view, Gates's introduction of Windows was tantamount to admitting that DOS was horrible software. After all, if DOS had been a good product, why would Microsoft have to hide it?

- As bad as DOS was, Windows was worse. Here's the evidence: By 1994, there were more than 35 products on the market that would allow users to cover Windows up to make it easier to use. Even Microsoft had a product, called "Bob," that was designed to make its "easy-to-use" Windows easy to use. Of course, in keeping with tradition, "Bob" was an even worse product than Windows.

- Which brings us to the latest phase in the evolution of mediocre products from Microsoft: Windows 95 and its successor Windows 98. Are they any good? Well, the revamped Windows products have been on the market for only a few years and rumor has it they are already being phased out.

- Judging by Bill Gates's wealth and the success of his company Microsoft, its easy to say that he's a marketing genius. However, judging the success of his products by the standards of science ("Does Windows answer questions about the universe that have heretofore gone unanswered?") or by the standards of technology ("Does Windows provide users with a tool that is better/faster/simpler than any tools available to them before?") makes it obvious that when it comes to science and technology, it is just as easy to call him a dismal failure.

Ever wonder how evolving technology evolves? In *Connections,* author James Burke looks at eight significant inventions, from the production line to plastics to the telecommunications to the television to the guided bomb to the computer, and traces their roots back to their humble—and surprising—beginnings. For a glimpse at how evolving science evolves, see Burke's *The Day the Universe Changed.*

HOW TO WIN THE NOBEL PRIZE . . .

Step 1

First, you have to do something noteworthy. The best way to do this is to devote your entire life to some intriguing question of science and spend every waking moment trying to get to the bottom of it. Keep in mind, however, the criteria by which Nobel Prize winners are judged: They must accomplish something that confers "the greatest benefit on mankind." (Ironically, the man who founded the Nobel Prize—Alfred Nobel—was the inventor of dynamite, an invention greatly prized in war.)

Step 2

Next, you have to get nominated. To do this, make friends with or otherwise impress lots of scientists, researchers, and university professors—because they are the ones who do the nominating. Since each year's prize winner is chosen by the Royal Swedish Academy of Sciences, it would be wise to focus your attention on the following groups in particular: Swedish and foreign members of the Academy of Sciences; permanent and assistant professors of physics at universities and institutes of technology in Sweden, Denmark, Finland, Iceland, and Norway; and professors at the Karolinska Institute, Sweden's only medical university. Oh, and it wouldn't be a bad idea to cozy up to other Nobel Prize winners.

Step 3

Wait. Once the Nobel committee receives the nominations, it must wade through them and pick out favorites. Then a vote is taken. If you're one of the lucky few, your name will be announced immediately after the vote is complete, sometime in October. You will then be flown to Stockholm for a fun-filled "Nobel week" and awarded your Nobel Prize on December 10 (which happens to be the anniversary of Alfred Nobel's death).

. . . AND WHAT TO DO AFTER YOU'VE WON IT
Physicists as Best-Selling Authors

Sometimes physicists write books that people like to buy (like *A Brief History of Time* by Stephen Hawking, even though he is not a Nobel Prize winner) and sometimes physicists write books that people like to read (like Richard Feynman's two autobiographical books, *Surely You're Joking, Mr. Feynman* and *What do YOU Care What Other People Think?*).

We've mentioned Feynman here often for two reasons: first, because he was a very funny guy, and second, because he had a way of explaining things that made the whole process of learning very enjoyable—whether he was talking about a phenomenon of physics or how to meet women in a bar or how to crack a safe.

THE ONE BOOK I WOULD READ . . .

If you're starting from scratch, I would suggest the *Encyclopedia Britannica*—because no matter what topic in physics you look up, at the end of the article they'll suggest other topics to look up that are related. For example, the eleventh edition, published in 1911, has an absolutely magnificent treatment of perpetual motion . . . it's a famous article.

With respect to a student who has some knowledge and would like to continue, I would recommend Richard Feynman's lectures—it's a three-volume set devoted to introductory physics called *The Feynman Lectures on Physics.*

—Val Fitch, winner of the Nobel Prize in physics in 1980

. . . when I was a small boy my father used to sit me on his lap and read to me from the *Encyclopedia Britannica*. . . . Our *Britannica* was the thirteenth edition—the same as the eleventh but made after the war—which was a famous edition that had good articles by very famous men. . . . the articles in the *Britannica* were written by such high-class people that they explained everything; the only problem was that there was so much, but everything was there if you worked at it.

—Richard Feynman, winner of the Nobel Prize in physics in 1965

. . . when I married, the *Encyclopedia Britannica* was Richard's wedding present to me—a house without a *Britannica* just didn't seem right to him.

—Joan Feynman, Richard's kid sister and fellow physicist

Hawking, as smart as he may be, isn't all that much fun to read. Yet, ironically, lots more members of the general public have bought his books than have bought Feynman's. On the other hand, we'd be willing to bet that a lot more people have read Feynman's books than have read Hawking's. Alas, the guys in marketing who're in charge of coffee-table books win again.

In addition to Feynman's autobiographical books, there's *No Ordinary Genius: The Illustrated Richard Feynman,* in which author and documentary filmmaker Christopher Sykes did a delightful job of bringing to life both Feynman's playful spirit and his brilliant science.

BEST-SELLING AUTHORS AS PHYSICISTS

Can *Dilbert* creator Scott Adams save the world?

In a recent magazine article, Adams claimed that he was devoting a healthy portion of his *Dilbert*-made fortune to fight malnutrition and world hunger by inventing a food that would be cheap, tasty, and easy to prepare and would provide all the nutrition you need for an entire day.

Why do we need this food? Said Adams:

> For different reasons. In my case because I'm really busy and it's just too damn difficult to find nutritional stuff that's quick to make and well-balanced—and in other people's cases because they're ignorant or they can't afford it, but in any case we're all going without the right foods. But there's no significant reason why we wouldn't have it all—except that nobody's working on it—it's nobody's job.
>
> So I've made it my job.
>
> I bought a Cuisinart and tried to whip together something that was kind of a burrito-looking thing. It had everything that I could best guess would be the most nutritionally balanced, and I wanted to see if I could get it to go together—but it was a complete failure because it tasted like crap and it didn't hang together very well.
>
> But it's my long-term goal to make some sort of food—maybe it's a burrito or maybe it's something else—but it will be some sort of food that would be fast to make, cheap to buy, and would contain just the right stuff to give you perfect nutrition for the day. This is my goal, whether it happens this year or fifty years from now.

A FINAL FEW WORDS

If you are not confused by quantum physics then you haven't really understood it.

—Niels Bohr

When we speak of the picture of nature in the exact science of our age, we do not mean a picture of nature so much as a picture of our relationships with nature.

—Werner Heisenberg

The outside world is something independent from man, something absolute, and the quest for the laws which apply to this absolute appeared to me as the most sublime scientific pursuit in life.

—Max Planck

Nothing in life is to be feared. It is only to be understood.

—Marie Curie

You never know when an old calendar might come in handy! Sure, it's not 1985 right now, but who knows what tomorrow will bring?

—Homer Simpson, on time and relativity

Index